T0192249

Introduction to Public Key Infrastructures

Johannes A. Buchmann • Evangelos Karatsiolis
Alexander Wiesmaier

Introduction to Public Key Infrastructures

Johannes A. Buchmann
FB Informatik
TU Darmstadt
Darmstadt
Germany

Evangelos Karatsiolis
FlexSecure GmbH
Darmstadt
Germany

Alexander Wiesmaier
AGT International
Darmstadt
Germany

ISBN 978-3-662-52450-3 ISBN 978-3-642-40657-7 (eBook)
DOI 10.1007/978-3-642-40657-7
Springer Heidelberg New York Dordrecht London

Softcover reprint of the hardcover 1st edition 2013

Printed on acid-free paper

Springer is part of Springer Science+Business Media (www.springer.com)

Preface

More than 30 years ago, when the Internet was emerging, public key cryptography was invented. Traditionally, cryptography relied on the exchange of secret keys prior to any secure communication, which made the application of cryptography in open networks such as the Internet very difficult. In contrast, public key cryptography allows for secure communication of entities that had no prior contact. Today, as the Internet has over two billion participants, this is extremely important. In addition, public key cryptography enables techniques that have no analogue in traditional cryptography, most importantly digital signatures. In fact, security on the Internet could not be achieved without digital signatures as they are, for example, required to authenticate software downloads and updates. We are convinced that today and in the future, there is and will be no IT security without public key cryptography.

Although public key cryptography does not rely on the exchange of secret keys, proper key management is still of vital importance to its security. In public key cryptography, pairs of private and public keys are used. The first task of such key management is to keep private keys private. This is easier than protecting keys in traditional secret key cryptography as there is no need to exchange private keys over insecure channels. But it is still an important challenge since there are billions of computing devices with private keys stored on them. The second task is to guarantee the authenticity of public keys, which is as important as maintaining the secrecy of private keys. For example, if the public signature verification key of a software vendor could be replaced by the public key of an adversary, the software signatures would be of no use since the adversary would be able to sign software in the name of the software vendor.

In order to fully understand public key cryptography, we therefore consider it necessary to study the infrastructures that manage key pairs in public key cryptography, the so-called public key infrastructures (PKIs). It is not sufficient to understand the ingenious mathematical mechanisms that underlie public key cryptography. This book grew out of a PKI course at Technische Universität Darmstadt, Germany, which we have been teaching for several years and which complements the introductory course on cryptography. It is our goal to cover the important concepts underlying PKI and to discuss relevant standards, implementations, and applications. We have included several exercises in each chapter that help deepen the understanding of its content. The book can thus be used as the basis for a course on PKI and as a self-study book for students and others interested in PKI. Only basic computer science knowledge is required. By giving numerous references that point to the relevant standards and implementation guidelines, we hope to make the book useful for those who are involved in PKI projects.

While writing this book and working on PKI projects, it became clear to us that PKI is still a very important research and development area. While public key cryptography applications that do not require user interaction are widely used (e.g., code signing), security solutions that require users to be actively involved are not so widespread (e.g., email signature and encryption). Many say that this is because current PKI concepts are still too complicated. Also, in the recent past, several incidents have shown that PKI does not always deliver the required security. Therefore, PKI concepts are required that overcome these deficiencies. We also intend this book to aid researchers and developers in doing so.

We would not have been able to write this book without the help of many people, in particular the students who attended the PKI course that the book is based on. Johannes Braun, Martin A. Gagliotti Vigil, Patrick Schmidt, Marcus Lippert, and Ciaran Mullan helped develop the exercises and made many important comments. We also thank Ronan Nugent and Alfred Hofmann of Springer for their support.

Darmstadt, Germany
July 2013

Johannes A. Buchmann
Evangelos Karatsiolis
Alexander Wiesmaier

Contents

Acronyms

AA	Attribute Authority
ACL	Access Control List
AES	Advanced Encryption Standard
APDU	Application Protocol Data Unit
API	Application Programming Interface
ARL	Authority Revocation List
ASN	Abstract Syntax Notation
AKI	Authority Key Identifier
BER	Basic Encoding Rules
CA	Certification Authority
CC	Common Criteria
CD	Compact Disc
CER	Canonical Encoding Rules
CMC	Certificate Management Messages over CMS
CMP	Certificate Management Protocol
CMS	Cryptographic Message Syntax
CPS	Certification Practice Statement
CRL	Certificate Revocation List
CRMF	Certificate Request Message Format
CSP	Certification Service Provider
CSP	Cryptographic Service Provider
CT-API	Card Terminal Application Programming Interface
CVC	Card Verifiable Certificate
DER	Distinguished Encoding Rules
DES	Data Encryption Standard
DIT	Directory Information Tree
DN	Distinguished Name
DNS	Domain Name System
DSA	Digital Signature Algorithm
DVD	Digital Video Disc
EBCA	European Bridge CA

ECB	Electronic Code Book
ECDSA	Elliptic Curve Digital Signature Algorithm
EEPROM	Electrically Erasable Programmable Read Only Memory
EFS	Encrypting File System
EV	Extended Validation
FINREAD	Financial Transactional IC Card Reader
FIPS	Federal Information Processing Standard
FTP	File Transfer Protocol
GNU	GNU's Not Unix
GPG	Gnu Privacy Guard
GSM	Global System for Mobile Communications
HSM	Hardware Security Module
HTTP	Hypertext Transfer Protocol
IBE	Identity-Based Encryption
ICC	Integrated Circuit Card
ICT	Information and Communication Technology
IETF	Internet Engineering Task Force
IP	Internet Protocol
ISP	Internet Service Provider
ITSEC	Information Technology Security Evaluation Criteria
ITU	International Telecommunication Union
JCRE	Java Card Runtime Environment
JCA	Java Cryptography Architecture
JCE	Java Cryptography Extension
JCEKS	Java Cryptography Extension KeyStore
JKS	Java KeyStore
LAN	Local Area Network
LDAP	Lightweight Directory Access Protocol
LRA	Local Registration Authority
MAC	Message Authentication Code
MIME	Multipurpose Internet Mail Extensions
OC	Object Class
OCF	Open Card Framework
OCSP	Online Certificate Status Protocol
OID	Object Identifier
OS	Operating System
PAM	Pluggable Authentication Module
PCI	Peripheral Component Interconnect
PC	Personal Computer
PC/SC	Personal Computer/Smart Card
PEM	Privacy Enhanced Mail
PER	Packed Encoding Rules
PGP	Pretty Good Privacy
PIN	Personal Identification Number
PKCS	Public Key Cryptography Standards

PKI	Public Key Infrastructure
PKC	Public Key Cryptography
PMI	Privilege Management Infrastructure
PoP	Proof of Possession
PSE	Personal Security Environment
PRNG	Pseudorandom Number Generator
PUK	Personal Unblocking Key
RA	Registration Authority
RDN	Relative Distinguished Name
RFC	Request for Comments
ROM	Read Only Memory
SASL	Simple Authentication and Security Layer
SCVP	Server-Based Certificate Validation Protocol
SHA	Secure Hash Algorithm
SIM	Subscriber Information Module
SKI	Subject Key Identifier
SMTP	Simple Mail Transfer Protocol
SPKI	Simple Public Key Infrastructure
SSH	Secure Shell
SSL	Secure Socket Layer
TBS	To Be Signed
TCP	Transmission Control Protocol
TCP/IP	Transmission Control Protocol/Internet Protocol
TSA	Time-Stamping Authority
TSL	Trust-Service Status List
TSP	Time-Stamp Protocol
TLS	Transport Layer Security
TOE	Target of Evaluation
URI	Uniform Resource Identifier
USB	Universal Serial Bus
UML	Unified Modeling Language
VPN	Virtual Private Network
W3C	World Wide Web Consortium
WAP	Wireless Application Protocol
WebDAV	Web-Based Distributed Authoring and Versioning
WLAN	Wireless Local Area Network
XER	XML Encoding Rules
X-KISS	XML Key Information Service Specification
XKMS	XML Key Management Specification
X-KRSS	XML Key Registration Service Specification
XML	Extensible Markup Language

Chapter 1
The Purpose of PKI

Public key cryptography helps make information and communication technology (ICT) systems more secure. Public key infrastructures (PKIs) enable the use of public key cryptography in open computer networks, in particular on the Internet. In this chapter, we use characteristic examples of Internet applications to illustrate potential threats against ICT and describe important security goals that can be achieved using public key cryptography and PKI. We discuss the basics of public key cryptography and explain how this technology supports the security goals explained in this chapter. Finally, we present the challenges that lead to the need for PKI. For more details on public key cryptography see [4].

1.1 The Internet

The Internet is the most important application domain of public key cryptography and public key infrastructures. It is an open "network of networks", i.e. a worldwide, publicly accessible network of interconnected computer networks that consists of millions of smaller domestic, academic, business, and government networks. It carries information and services, such as electronic mail, file transfer, social networks and all kinds of other Web services. Nowadays, everything from health care to business, production, and recreation activities depends on computers. Most of these computers have access to the Internet. Thus, potentially anyone in the world can access them. As of June 2012, 2.41 billion people in the world used the Internet [8]. This is 34.3 % of the world population and a growth of 566.4 % since 2000.

An important Internet application is *electronic commerce* which is also called *e-commerce* for short. E-commerce refers to any type of business on the Internet. This covers a broad range: from Internet-based retail sites, through auction and music sites, to exchanges of trading goods and services between corporations. E-commerce allows consumers to electronically exchange goods and services without barriers of time or distance. E-commerce has expanded rapidly recently. The boundaries between "conventional" and "electronic" commerce have become

J.A. Buchmann et al., *Introduction to Public Key Infrastructures*,
DOI 10.1007/978-3-642-40657-7_1,

increasingly blurred as more and more businesses move their operations to the Internet.

Another Internet application is *e-health*. E-health encompasses services such as maintenance of electronic medical records that enable easy communication of patient data between different health care professionals, organizations, and businesses. Another example of e-health is telemedicine, which includes all types of physical and psychological measurements that do not require a patient to travel to a specialist. Also, e-health enables virtual health care teams which consist of health care professionals who collaborate and share information on patients through digital equipment for transmural care.

Electronic government or shortly *e-government*, our third example, refers to governmental use of the Internet to exchange information and services with citizens, businesses, and other arms of government. E-government is applied by many governmental units such as legislature, judiciary, or administration, in order to improve internal efficiency and the delivery, accessibility, and convenience of public services, or processes of democratic governance.

1.2 Security Goals

Many of the Internet services mentioned in Sect. 1.1 are highly security-sensitive. This was emphasized in a white paper of the US Clinton administration from 1998 [20]. "The United States possesses both the world's strongest military and its largest national economy.... Because of our military strength, future enemies, whether nations, groups or individuals, may seek to harm us in nontraditional ways.... Our economy is increasingly reliant upon interdependent and cyber-supported infrastructures and nontraditional attacks on our infrastructures and information systems may be capable of significantly harming both our military power and our economy." As a consequence, the Internet must be protected from malicious attacks. In this chapter we describe in more detail the protection goals confidentiality, integrity, entity authentication, data authenticity, and non-repudiation that can be achieved using public key cryptography.

1.2.1 Confidentiality

A very important security goal is *confidentiality*. It refers to the property that data or information is not made available or disclosed to unauthorized persons or processes. Confidentiality is closely related to *privacy*. Privacy is the ability of an individual or group to keep information about itself confidential and to control the access to and use of such information.

We present a few examples where confidentiality is an important protection goal.

As early as 1977 Martin Gardner wrote [7]: "Government agencies and large businesses will presumably be the first to make extensive use of electronic mail, followed by small businesses and private individuals. When this starts to happen, it will become increasingly desirable to have fast, efficient ciphers to safeguard information from electronic eavesdroppers." Today, email is widely used for private, business, and government communication. In many cases, emails contain confidential information. But most email messages can be easily intercepted. They travel through an indeterminate set of systems and network devices, each of which offers a point of interception. These systems are operated, for example, by companies, universities, governments, or telecommunication providers. In such environments adversaries may easily gain access to email messages.

Confidentiality and privacy is also of great importance in the context of e-health. Unauthorized persons and processes must be prevented from seeing patient data that is communicated between different health care professionals, organizations, and businesses. Likewise, virtual health care teams that share information on patients must keep this information confidential.

Maintaining confidentiality is also a serious issue in e-government. On 21 November 2007, *The New York Times* reported: "The British government struggled Wednesday to explain its loss of computer disks containing detailed personal information on 25 million Britons, including an unknown number of bank account identifiers, in what analysts described as potentially the most significant privacy breach of the digital era" [15].

1.2.2 Integrity

Integrity refers to the property that data has not been modified. There are many contexts in which being able to check the integrity is very important. For example, the integrity of software running in a security-sensitive context such as e-health, e-commerce, or e-government must be protected in order to prevent malicious changes. Such changes may for example cause the systems to transmit confidential information to unauthorized entities. Since integrity is such an important protection goal, common criteria (CC) evaluations require evaluated software, the so-called target of evaluation (TOE), to provide "controls to ensure that unauthorized modifications are not made to the TOE" [5]. Examples of devices that are required to be CC-evaluated are the device that connects a medical practice in Germany to the Internet, the so-called *Konnektor*, and the microprocessors that are used in European electronic passports.

1.2.3 Entity Authentication

Entity authentication is a process by which a *verifier* is assured of the identity of a *prover* whose identity is to be established. Instead of entity authentication we also use the term *identification*. Both the verifier and the prover can be persons,

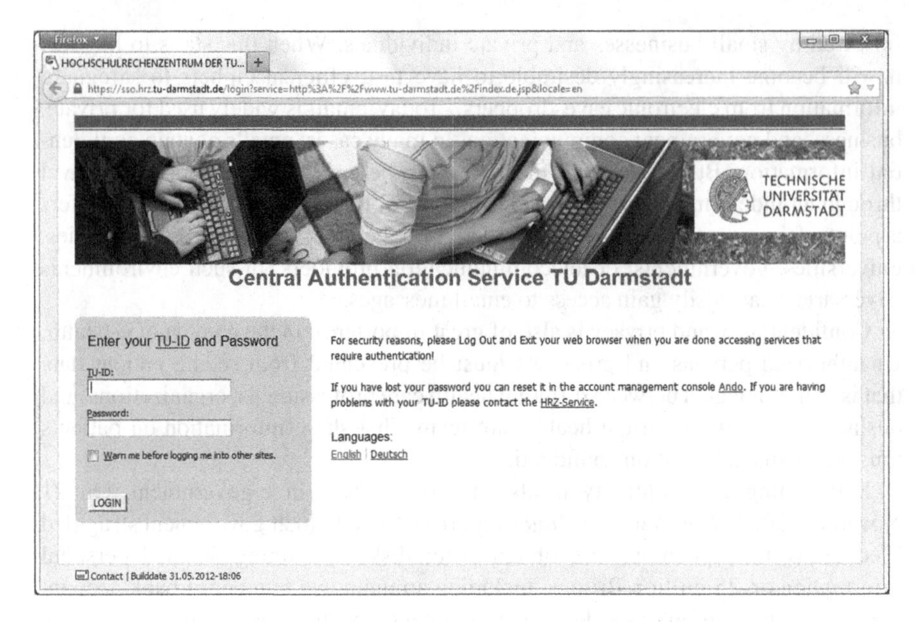

Fig. 1.1 The login process of TU Darmstadt

IT components, IT processes, and so on. Entity authentication is different from data authenticity, which is described in Sect. 1.2.4. Entity authentication is required by many applications, for example, in the online authentication system of TU Darmstadt, which is shown in Fig. 1.1. First, users register with TU Darmstadt by providing their name and selecting a secret password. When users want to log in later to their TU Darmstadt account they submit this user name and password.

Identification with user name and password is quite common in e-commerce, e-government, and e-health applications. However, there are rather effective attacks against this identification method. An example is that of *phishing* attacks. In such an attack, the adversary attempts to fraudulently acquire sensitive information, such as user names, passwords and credit card details, by masquerading as a trustworthy entity in an electronic communication. As we will see in Sect. 1.3.7, PKI provides stronger entity authentication mechanisms.

1.2.4 Data Authenticity

Authenticity of data includes integrity and additionally requires that the origin of the data can be determined.

Authenticity of data is required in many contexts and applications. For example, banks must be certain that electronic financial transactions are authentic. Likewise, security-sensitive emails must be authentic. Another example is software updates. Such updates become necessary when software errors, in particular security holes in operating systems and application software, are discovered or newer versions with more functionality become available. Updating computers must authenticate all updates. Otherwise, an adversary may be able to supply updates that cause severe damage. For example, a fraudulent operating system update may erase the computer's hard disc, thereby destroying all data of all users of that computer. Remote updates are also used for smart phones and many other devices.

1.2.5 Non-repudiation

Non-repudiation is a property of data or processes which prevents an entity from denying having performed a particular action, for example having sent an email or having initiated a money transfer.

As an example, consider electronic contracts. In many countries they are as binding as paper contracts. The parties involved in such a contract must not be able to repudiate this contract later. Also, physicians who issue electronic prescriptions must not be able to repudiate the prescription later. Likewise, electronic tax declarations must not be repudiable at a later point in time.

1.2.6 Other Security Goals

There are many more security goals in which PKI plays an important role. An example is *anonymity*. It is a condition in which an individual's true identity is unknown. For example, anonymity is important in electronic elections where the voters have to remain anonymous. Another security goal which is relevant in the context of electronic elections is *uncoercibility*. Coercibility means that an adversary can force a voter to cast a certain vote.

1.3 Cryptography

In the previous section we have described important security goals that arise in the electronic world. In this section, we review basic mechanisms from cryptography and specifically from public key cryptography. We explain how these techniques help achieving the described security goals.

1.3.1 Secret Key Encryption

Traditionally, confidentiality of data is provided by secret key cryptosystems. Examples of such cryptosystems are the Data Encryption Standard (DES) [11],

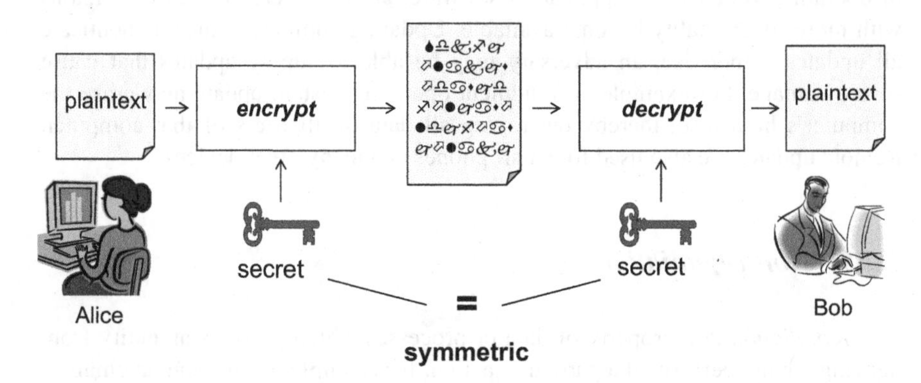

Fig. 1.2 The process of secret key encryption

Table 1.1 Performance of symmetric cryptosystems

AES	Twofish	RC6	Serpent	DES-ede
35 ms	31 ms	20 ms	58 ms	141 ms

which was selected as an official Federal Information Processing Standard (FIPS) for the United States in 1976 and published in 1977, and its successor, the Advanced Encryption Standard (AES) [14]. AES became effective as a standard in 2002. DES in its original form is no longer secure. However, there are still secure variants of DES such as DES-ede [12]. There are many more secret key encryption schemes, for example RC6 [16], Serpent [18], and Twofish [22].

In a secret key cryptosystem, the communication partners—call them Alice and Bob—agree on a secret key before they secretly communicate. For this *key agreement* they may use a secure channel, for example a courier. Alice and Bob can also use the Diffie-Hellman key exchange protocol [6]. It was invented in 1976 and does not require a secure channel.

Figure 1.2 shows the process of secret key encryption. Suppose that Alice and Bob have successfully exchanged a secret key. If Alice wishes to send a confidential message to Bob, she uses this secret key to encrypt the data to be kept confidential. The data that is to be encrypted is called the *plaintext*. The result of the encryption is the *ciphertext*. Alice sends the ciphertext to Bob. Upon receiving the ciphertext, Bob uses the secret key that he has exchanged with Alice to decrypt the ciphertext. If the cryptosystem is secure, then nobody can obtain any information concerning the plaintext without the knowledge of the secret key.

Secret key cryptosystems are also referred to as *symmetric* cryptosystems since both Alice and Bob have the same key. In some symmetric encryption schemes, the encryption and decryption keys are different. However, they can be easily computed from each other. Some secret key cryptosystems that are used in practice and the time required to encrypt 1 MByte on a dual core 2.8 GHz computer using the Java library Bouncy Castle [21] are shown in Table 1.1. All algorithms have a block size of 128 bits, except DES-ede, which has the block size 64. The mode of use is the electronic code book (ECB) mode.

Table 1.2 A public key directory

Name	Public key
Buchmann	13121311235912753192375134123
Karatsiolis	84228349645098236102631135768
Wiesmaier	73528720920032836216521830930
Alice	54628291982624638121025032510
Bob	27381253812351972497652990930
⋮	⋮

1.3.2 *Public Key Encryption*

With the development of open computer networks such as the Internet with its billions of users, exchanging secret keys between potential communication partners as required by secret key cryptography became rather impractical.

One possible solution of this key distribution problem is to let a central authority, the *key center*, distribute secret keys to all users. Applying these keys, the users can secretly communicate with each other. An example of such an architecture is the mobile phone standard GSM. In fact, GSM uses many interconnected key centers. Each GSM provider operates one or more key centers. However, key distribution by a centralized key center has a serious disadvantage. The key center can potentially access all secret messages. This only works in closed environments where all users trust the central authorities, for example their mobile phone providers. In an open, decentralized environment such as the Internet, key distribution by key centers is inappropriate. For such a situation, public key cryptography was invented.

The main idea of a *public key cryptosystem* is that two different but related keys are used: one for encryption and one for decryption. Decryption works with the decryption key only and not with the encryption key, and vice versa. As the decryption key cannot be determined from the encryption key, the encryption key can be made public. In order for a user Bob to receive confidential messages, he uses a key pair consisting of an encryption key and the corresponding decryption key. Bob keeps the decryption key secret while his encryption key is made public (as illustrated in Table 1.2). The encryption key is called Bob's *public key*; the corresponding decryption key is called Bob's *private key*. Once the keys are generated and Bob's public key is published, Bob can receive confidential messages from anyone. No further key distribution is required.

This is illustrated in Fig. 1.3. Alice wishes to send a confidential message to Bob. She uses Bob's public key to encrypt the message. The result of this encryption is a ciphertext which is sent to Bob. Bob decrypts this message using his private key. Since Bob keeps his private key secret, Alice's message remains confidential. This shows that in public key cryptosystems key distribution only requires making public keys accessible.

Public key cryptosystems are also called *asymmetric cryptosystems*. Public key encryption not only provides confidentiality. It can also be used to implement identification protocols. Suppose that the prover Bob who possesses a secret

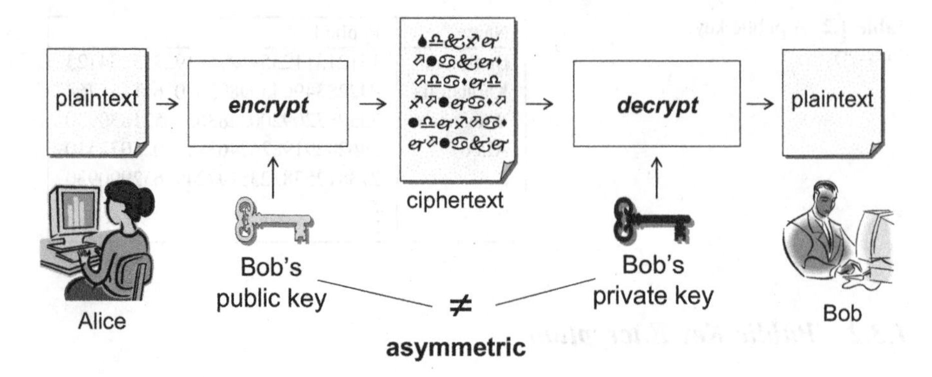

Fig. 1.3 The process of asymmetric encryption

encryption key wishes to identify himself to the verifier Alice. Alice selects a random number, encrypts it with Bob's public key and sends the ciphertext to Bob. Upon receiving the ciphertext, Bob decrypts it and sends the random number back to Alice. Alice is convinced of Bob's identity if the random number that she receives from Bob is the same as the random number that she has sent to him. However, this simple protocol has several drawbacks. One problem is that Alice can make Bob decrypt messages of her choice since Bob may not be able to distinguish encrypted random numbers from other ciphertexts. For this reason, public key encryption is typically not used for entity authentication. Instead, digital signatures are used, these are described in Sect. 1.3.7.

1.3.3 The RSA Public Key Cryptosystem

The first and most frequently used public key cryptosystem is the *RSA cryptosystem*. It is named after its inventors Rivest, Shamir, and Adleman, who received the 2002 Turing award for their invention. We explain a basic version of the RSA cryptosystem. For details see [4].

To generate his secret and the corresponding public key, Bob selects two large prime numbers p and q and computes their product

$$n = pq. \tag{1.1}$$

Bob also chooses an integer e with

$$1 < e < \varphi(n) = (p-1)(q-1) \text{ and } \gcd(e,(p-1)(q-1)) = 1. \tag{1.2}$$

Note that e is always odd since $(p-1)(q-1)$ is even. Bob computes an integer d with

$$1 < d < (p-1)(q-1) \text{ and } de \equiv 1 \bmod (p-1)(q-1). \tag{1.3}$$

That number can be computed by the extended Euclidean algorithm. Bob's public key is the pair (n, e). His private key is d. The number n is called the *RSA modulus*, e is called the *encryption exponent*, and d is called the *decryption exponent*. Note that the private key d can be computed from the encryption exponent e if the prime factors p and q of the RSA modulus n are known. Therefore, if an adversary—call her Eve—is able to find the prime factorization of n, then she can easily find Bob's private key d. In fact, it can be shown that finding the private key d requires the knowledge of the secret factors p and q.

Example 1.1. Bob chooses the prime factors $p = 11$ and $q = 23$. Then $n = 253$ and $(p-1)(q-1) = 10*22 = 220$. The smallest possible encryption exponent e is $e = 3$ since $\gcd(3, 220) = 1$. For this encryption exponent, the extended Euclidean algorithm yields the decryption exponent $d = 147$.

The possible plaintexts for the RSA cryptosystem are the integers m with

$$0 \leq m < n. \tag{1.4}$$

A plaintext m is encrypted by raising it to the eth power mod n. So the ciphertext is

$$c = m^e \bmod n. \tag{1.5}$$

If Alice knows the public key (n, e), she can encrypt m.

Example 1.2. As in Example 1.1, let $n = 253$ and $e = 3$. Then the plaintext space is $\{0, 1, \ldots, 252\}$. Encrypting the integer $m = 165$, Alice obtains $165^3 \bmod 253 = 110$.

To decrypt the ciphertext c, Bob computes the dth power of $c \bmod n$. So we have

$$m = c^d \bmod n. \tag{1.6}$$

Example 1.3. We conclude Examples 1.1 and 1.2. There, we have chosen $n = 253$, $e = 3$, and $d = 147$. Moreover, we have computed the ciphertext $c = 110$. Bob can reconstruct the plaintext as $110^{147} \bmod 253 = 165$.

RSA works because raising integers to eth powers mod n is easy for anybody who knows the public key (n, e). But for sufficiently large RSA moduli extracting eth roots mod n is intractable without knowledge of the private key d.

1.3.4 Other Public Key Cryptosystems

Algorithms for factoring composite integers has been a research topic for many decades. After the invention of RSA, this research was intensified but no efficient factoring algorithm has been found that works on classical computers. This does not mean that polynomial time factoring algorithms are impossible. On the contrary,

Table 1.3 Performance of public key cryptosystems

RSA	ElGamal
3.3 s	423 s

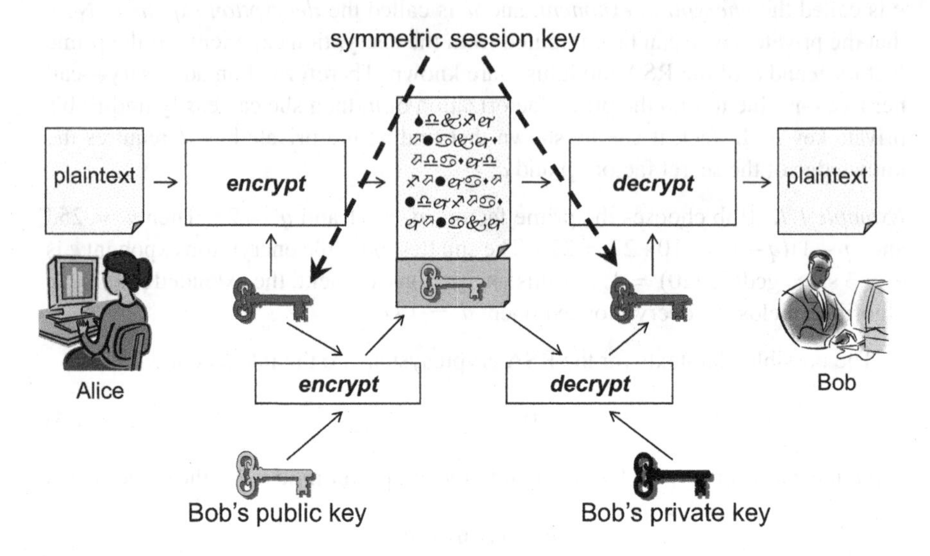

Fig. 1.4 Hybrid encryption

in 1994 Peter Shor invented a polynomial time factoring algorithm for quantum computers [19]. However, building sufficiently large quantum computers for factoring large numbers is still out of reach. In the future, there may also appear new efficient factoring algorithms. It is therefore not sufficient to rely on the RSA cryptosystem. Alternatives are, for example, elliptic curve cryptosystems. Their security is based on the difficulty of computing discrete logarithms in the group of points of an elliptic curve over finite fields. Elliptic curve cryptosystems use much smaller keys than RSA. However, discrete logarithm-based cryptosystems are also subject to efficient quantum computer attacks. An alternative cryptosystem that appears to be immune against quantum computer attacks is the McEliece cryptosystem (see [2]).

1.3.5 Hybrid Encryption

The known public key cryptosystems are not as efficient as many symmetric cryptosystems. Two public key cryptosystems and the time required to encrypt 1 MByte on a dual core 2.8 GHz computer using the Java library Bouncy Castle and the padding method described in the PKCS#1 standard [17] are shown in Table 1.3.

Therefore, in practice, *hybrid encryption* is used. Hybrid encryption is an efficient combination of public key and secret key cryptosystems. It is illustrated in Fig. 1.4. In principle, hybrid encryption works as follows.

Hybrid encryption uses an efficient symmetric cryptosystem together with a public key scheme. Alice wants to send an encrypted message to Bob. She generates a *session key* for the symmetric cryptosystem. At this point of the protocol, this session key is not known to Bob. It is only known to Alice who generated it. Alice encrypts the message for Bob using the symmetric encryption scheme and the session key. She obtains a ciphertext. Since the symmetric cryptosystem is efficient, this encryption is fast. But Bob is not able to decrypt the ciphertext as long as he does not know the session key. Therefore, Alice encrypts the session key with Bob's public key. Since the session key is small, this encryption is also fast although the encryption function of the public key system may not be very efficient. Alice sends the ciphertext and the encrypted session key to Bob. Bob decrypts the session key using his private key. Then he decrypts the ciphertext using the session key and obtains the original message. Here, the public key system is only used for the confidential transport of the session key. This is how hybrid encryption combines the elegance of key management of public key cryptosystems with the efficiency of secret key cryptosystems.

1.3.6 Cryptographic Hash Functions and Message Authentication Codes

Hash functions map arbitrary bit strings to strings of fixed length. They are of the form

$$h : \{0, 1\}^* \rightarrow \{0, 1\}^n \tag{1.7}$$

where n is a positive integer. For simplicity, we have used the alphabet $\{0, 1\}$, which may be replaced by any other finite set. Cryptographic hash functions are hash functions with additional security properties. The most important property in our context is *collision-resistance*. The hash function (1.7) is called collision-resistant if finding two distinct strings w and w' with the same hash value $h(w) = h(w')$ is infeasible.

Cryptographic hash functions may be used to establish integrity, for example of software. When software s is installed on some computing device, its hash value $x = h(s)$ is calculated and stored in some other secure place, for example on a smart card. When it comes to checking the integrity of the software s at a later point in time, the hash value $x' = h(s')$ of the potentially modified software s' is calculated. If the two hash values $h(s)$ and $h(s')$ are identical, then the integrity of the software is proved. This follows from the fact that the hash function is collision-resistant.

In order to provide data authenticity, a *message authentication code* (MAC) can be used. MACs are closely related to cryptographic hash functions. A MAC works as follows. If Alice wishes to exchange authenticated messages with Bob they first agree on a secret MAC key. To send an authenticated message to Bob, Alice uses this key to compute the MAC of the message. Like hash values, MACs are also short

Table 1.4 Performance of MAC algorithms

HMAC-SHA1	HMAC-SHA512	HMAC-MD5	HMAC-Tiger
16 ms	53 ms	9 ms	24 ms

bit strings of fixed length. Alice sends the message together with the MAC to Bob. Bob can use the shared secret key to verify the MAC.

In Table 1.4 we present a few MAC algorithms and show their performance on processing of 1 MByte on a dual core 2.8 GHz using the Java library Bouncy Castle.

MACs can in principle also be used for entity authentication. Alice and Bob agree on a secret key. When Alice wants to authenticate Bob, she sends a random string to Bob. Bob calculates the MAC of that string and returns it to Alice. Alice verifies Bob's identity by verifying the MAC. However, this identification procedure has the disadvantage of only working between two parties. Bob cannot use the same secret key to authenticate himself to other communication partners since Alice, knowing the secret, could impersonate Bob. This problem can be solved if an authentication center is used that shares a secret key with all users. In order for Alice to identify Bob she delegates the identification of Bob to this authentication center.

Identification solutions that do not rely on authentication centers are based on digital signatures or dedicated identification protocols, which are discussed in the following sections.

1.3.7 Digital Signatures

In many contexts, entity and data authentication cannot be based on symmetric techniques such as MACs. As an example, consider software updates. If a software manufacturer such as Adobe or Microsoft issues a software update, it is important for the users to be able to authenticate this update. This has been explained in Sect. 1.2.4. However, the users must not be able to impersonate the software vendor and distribute their own software updates. Therefore, MACs cannot be used in this context. An asymmetric authentication mechanism is required that allows computers to authenticate software updates without being able to generate authentic software updates themselves. *Digital* or *electronic signature schemes* have this property.

In a digital signature scheme, the *signer* Bob uses his secret *signature key* to calculate digital signatures of documents. That key is also referred to as Bob's *private key*. Potential verifiers can use Bob's public *verification key* that corresponds to the secret signature key to verify Bob's digital signature. The verification key is also called Bob's *public key*. As in public key cryptosystems, the private keys in a signature scheme cannot be computed from the corresponding public keys. Figure 1.5 illustrates digital signatures.

Digital signatures can be used to prove the integrity and authenticity of data. Also, digital signatures provide entity authentication and non-repudiation. Because of those numerous applications, digital signatures are an extremely important cryptographic tool. We now explain the various uses of digital signatures in more detail.

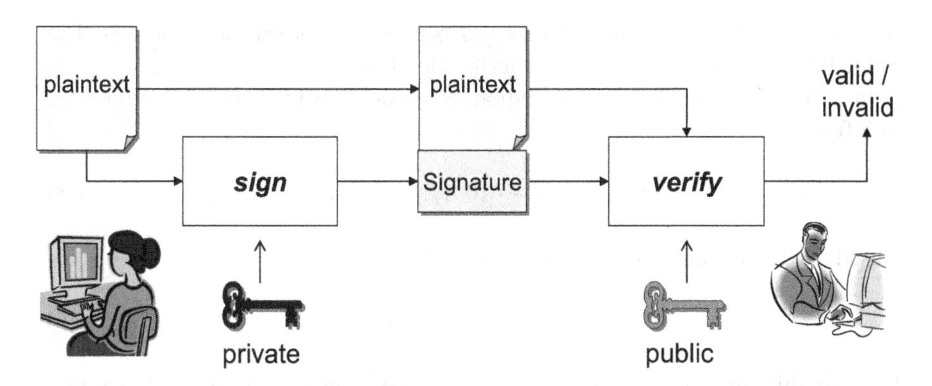

Fig. 1.5 Digital signatures

Suppose that Bob signs data such as an email or a bank transaction digitally. By verifying this signature, Alice convinces herself that the data origin is Bob and that the data has not been altered. This way, integrity and authenticity of the data are established. In fact, many software vendors such as Microsoft, Apple, and Adobe use digital signatures to authenticate software updates.

Next we explain that digital signatures also provide non-repudiation. Alice and Bob agree on a contract. They both sign this contract digitally. At a later point in time, Bob repudiates the contract. Alice presents Bob's digital signature on the contract to a judge in order to prove that Bob agreed to the contract. The judge obtains Bob's public key and verifies the signature. It convinces the judge that Bob has in fact signed the contract.

Digital signatures can also be used for entity authentication. There are several ways to realize this. We explain a basic *challenge-response identification scheme* that uses digital signatures. Suppose that Alice wishes to identify herself to Bob. Bob sends a random number (the challenge) to Alice. She signs that random number and sends the digital signature (the response) to Bob. The successful verification of the signature will convince Bob of Alice's identity. Since there is a new random challenge each time, identification based on digital signatures is more secure than user name with password identification.

There are many variants of the identification protocol described here. For example, the SSL and TLS protocols combine identification with setting up a secure channel. This will be discussed in Chap. 10.

1.3.8 The RSA Signature Scheme

The most widely used digital signature algorithm is the RSA scheme. Key generation for this algorithm works exactly as in the RSA cryptosystem presented in Sect. 1.3.3. To describe the operations of signing and verifying we use the notation introduced in Sect. 1.3.3. The private key d becomes the private signature key. The public key (n, e) becomes the public verification key.

In addition to signature and verification keys, a publicly known collision-resistant hash function

$$h : \{0, 1\}^* \to \{0, 1, \ldots, n-1\} \tag{1.8}$$

is used. The signature of a string $D \in \{0, 1\}^n$ is

$$s = h(D)^d \bmod n. \tag{1.9}$$

To verify this signature, the verifier uses the signer's public key (n, e), determines the hash value $h(D)$ and checks that

$$h(D) = s^e \bmod n. \tag{1.10}$$

If this congruence holds, then the signature is valid. Otherwise, it is invalid.

Example 1.4. Alice chooses $p = 11$, $q = 23$, $e = 3$. She obtains $n = 253$, $d = 147$. Alice's public key is $(253, 3)$. Her private key is 147.

Alice wants to sign a data string D with hash value $h(D) = 111$. She computes $s = 111^{147} \bmod 253 = 89$. The verifier verifies the signature by computing $s^3 \bmod 253 = 111$ and checking that this value matches with the hash value.

It is also possible to generate the signature as

$$s = D^d \bmod n \tag{1.11}$$

without previously hashing the document to be signed. Then the document has to be chosen from the set $\{0, 1, \ldots, n-1\}$ and it is possible to recover the document by verifying the signature

$$D = s^e \bmod n. \tag{1.12}$$

This is called a *signature with message recovery*. As explained in [4], using RSA with message recovery opens the possibility of existential forgeries.

1.3.9 Other Digital Signature Schemes

There are several alternatives to the RSA signature scheme. Alternative systems that are used in practice are the digital signature algorithm (DSA) [13] and the elliptic curve digital signature algorithm (ECDSA) [9]. Their security is based on the difficulty of computing discrete logarithms in the multiplicative group of a finite field or in the group of points of an elliptic curve over a finite field, respectively.

As explained in Sect. 1.3.4, sufficiently large quantum computers can factor RSA moduli and compute discrete logarithms very efficiently. An alternative efficient signature scheme that appears to be quantum computer-immune is the Merkle Signature Scheme and its extensions [2].

1.4 Why Public Key Infrastructure?

We have seen how public key cryptography achieves important IT security goals. The advantage of public key cryptography is that no exchange of secret keys is required. But in public key cryptography, keys must also be managed appropriately. The efficient and secure management of the key pairs during their whole life cycle is the purpose of public key infrastructures. This life cycle is the following.

The first phase is *key generation*. In this step, a key pair is created. The next phase is *key usage*. In this phase, the private key is used to decrypt or sign data. Also, users have access to the public keys of other users to encrypt ciphertexts or to verify signatures. In the final stage of its life cycle a key pair becomes invalid. For example, this happens when the validity period of the key pair ends or when the key pair is invalidated because it is compromised.

There are several tasks of a PKI associated with each stage of the life cycle of key pairs.

The first task is to ensure that in the key generation phase, secure key pairs are produced. One option is to let the users generate their own key pairs. Then they can prevent the private keys from being disclosed to unauthorized persons. However, due to technical limitations, users may not be able to generate secure keys. For example, creating keys on computers that are connected to the Internet may be insecure since those computers may be infected by malware. Hence, it appears to be more secure to generate keys on dedicated hardware such as smart cards. However, this may also be problematic. For example, consider the RSA system. RSA keys are generated using two random prime numbers of approximately equal size. In the generation of those prime numbers pseudorandom number generators (PRNG) are used. Those PRNGs must be cryptographically secure, which means that predicting their result must be infeasible. This is necessary to prevent adversaries from reconstructing the private keys of users. But smart cards may not have enough computational resources to run secure PRNGs. An alternative is to have the key pairs generated by a trusted third party which has sufficient computational power to generate secure keys. However, if a trusted third party generates the key pairs, the private keys become also known to that party, which may compromise the confidentiality of private keys. Thus, establishing an appropriate process for key generation is a challenging task for a PKI.

In the key usage phase, a PKI also has several important tasks. The most important and complex task of a PKI is to *make the public keys available to the users*. It is not sufficient to simply publish the public keys. Those who use public keys must be able to verify their *authenticity* and *validity* and must know their *properties*. If the authenticity of Bob's public keys is not guaranteed, the adversary Oscar may be able to replace Bob's public keys with his own public keys. This may enable him to decrypt messages that were encrypted for Bob or sign documents in the name of Bob. Key validity refers to several issues. Public keys become invalid when they expire. Also, public keys may be intended for a certain purpose. For example, for security reasons, signature keys may be intended for document signing but not for identification. An example of an important property of a key pair is the policy that was applied when the key was generated. For example, this policy determines the level of protection that was used during key generation. Users may find this protection level too low and may therefore decide not to use the key.

The next task of a PKI is to *deal with the problem of public keys becoming insecure*. This may happen for various reasons. For example, if the smart card on which Alice keeps her secret signature key is stolen, the corresponding public key must no longer be used to verify signatures of Alice. PKI users must be informed that this is the case. Even more dramatically, if a public key cryptosystem is broken, then all public keys that have been issued using this public key cryptosystem must be invalidated.

Another task of a PKI in the key usage phase is to enable *key backup* whenever necessary. For example, if Alice loses her smart card with her private decryption key she can no longer access her encrypted files unless there is a backup of that decryption key. Since private keys must be kept confidential, key backup is a security-sensitive issue. Note that private signature keys do not require backups. When they are lost they may be replaced with new keys.

The final task of a PKI in the usage phase is to support the protection of private keys. This can be done by making secure software or hardware components such as smart cards available and managing those components.

Once keys expire or become invalid for some other reason, the PKI must *manage* those *invalid keys*. One option is to destroy them. Another option is to archive them for later use. Invalid public encryption keys may be deleted. However, expired private decryption keys may have to be archived to maintain the accessibility of data that was encrypted using the corresponding encryption keys. Likewise, invalid private signature keys may be deleted as they must no longer be used to sign data. However, public verification keys may have to be archived to allow verification of signatures that are used to provide long-term authentication or non-repudiation.

1.5 Identity-Based Public Key Cryptography

As we will see in the remainder of this book, setting up a PKI in which public keys are available for all users is a challenging task. In order for users to be able to access those public keys, they must be online or keep archives of relevant public keys. For some contexts this is inappropriate. For example, consider an emergency situation in which rescue workers, fire fighters, and police officers wish to communicate confidentially. Since they are an ad hoc group, they do not know each other's public keys. They also may not have access to a central public key directory. For such a group of users, *identity-based public key cryptography* is very useful. It allows them to compute the public keys from each other's identities. No further authentication of the public keys is needed.

Identity-based cryptography does not require a public key infrastructure that distributes keys. However, there are several issues that must be considered. In identity-based cryptography, the private keys of users are computed by a trusted authority that has a secret master key. In principle, that authority can use all these private keys. Also, revocation of invalid keys is more challenging as the public keys are computed from identities that cannot be revoked. Because of these deficiencies, identity-based public key cryptography is only used in niche applications.

In practice, mostly identity-based encryption (IBE) is used. In [3] the implementation guidelines for two identity-based encryption algorithms are presented. An infrastructure for IBE is described in [1]. Using IBE for sending encrypted emails is specified in [10].

1.6 Object Identifiers

To make cryptography interoperable, standards are necessary that fix sizes, parameters, formats, and so on. Standardized cryptographic algorithms are referred to by *object identifiers* (OIDs). Table 1.5 presents OIDs of the most important cryptosystems. For example, the OID "1.2.840.113549.1.1.5" stands for the signature scheme RSA used in combination with the cryptographic hash function SHA1.

OIDs identify arbitrary objects and not only cryptographic algorithms. Therefore they are used extensively in a PKI for interoperability purposes.

1.7 Exercises

1.1. A bookseller offers books for sale on the Internet. He publishes the list of offered books on his Web page. A customer selects books and specifies the payment method. Upon receiving the estimated delivery time, price, and delivery details the customer completes the order and sends the necessary data for payment and processing. The dealer replies with an acknowledgement. Which security goals are relevant in which step of the transaction?

1.2. Identification may be based on the proof of user characteristics of the following types:

- Properties (e.g. eye color).
- Abilities (e.g. a handwritten signature).
- Knowledge (e.g. a password).
- Possession (e.g. an entrance ticket).

Table 1.5 Examples of OIDs

Algorithm	Type	OID
MD5	Cryptographic hash function	1.2.840.113549.2.5
SHA1	Cryptographic hash function	1.3.14.3.2.26
SHA256	Cryptographic hash function	2.16.840.1.101.3.4.2.1
SHA384	Cryptographic hash function	2.16.840.1.101.3.4.2.2
SHA512	Cryptographic hash function	2.16.840.1.101.3.4.2.3
SHA256withDSA	Digital signature	2.16.840.1.101.3.4.3.2
SHA256withECDSA	Digital signature	1.2.840.10045.4.3.2
SHA384withECDSA	Digital signature	1.2.840.10045.4.3.3
SHA512withECDSA	Digital signature	1.2.840.10045.4.3.4
MD5withRSA	Digital signature	1.2.840.113549.1.1.4
SHA1withRSA	Digital signature	1.2.840.113549.1.1.5
SHA1withDSA	Digital signature	1.2.840.10040.4.3
SHA1withECDSA	Digital signature	1.2.840.10045.4.1
AES with 128 bit key in ECB mode	Secret key encryption	2.16.840.1.101.3.4.1.1
AES with 256 bit key in CBC mode	Secret key encryption	2.16.840.1.101.3.4.1.42
HMAC-MD5	MAC	1.3.6.1.5.5.8.1.1
HMAC-SHA1	MAC	1.3.6.1.5.5.8.1.2
RSA	Public key encryption	1.2.840.113549.1.1.1

Give further examples of each type.

1.3. In order to strengthen password identification, identification servers store cryptographic hash values of passwords. When a user identifies herself, she types in the password, the client computes and transmits the hash of the password, and the server compares the transmitted value with its stored hash value.

1. Discuss the security of this method. In particular, show how a replay attack can be mounted.
2. How can the replay attack be prevented?

References

1. G. Appenzeller, L. Martin, M. Schertler, Identity-based encryption architecture and supporting data structures, in *IETF Request for Comments*, 5408, Jan 2009
2. D.J. Bernstein, J. Buchmann, E. Dahmen (ed.), *Post Quantum Cryptography* (Springer, Berlin, 2008)
3. X. Boyen, L. Martin, Identity-based cryptography standard (IBCS) #1: supersingular curve implementations of the BF and BB1 cryptosystems, in *IETF Request for Comments*, 5091, Dec 2007
4. J.A. Buchmann, *Introduction to Cryptography*, 2nd edn. (Springer, New York, 2004)
5. Common Criteria, Common criteria for information technology security evaluation—part 3: security assurance components—version 3.1 (2007), http://www.commoncriteriaportal.org/thecc.html

6. W. Diffie, M.E. Hellman, New directions in cryptography. IEEE Trans. Inf. Theory IT-**22**(6), 644–654 (1976)
7. M. Gardner, Mathematical games: a new kind of cipher that would take millions of years to break. Sci. Am. **237**(2), 120–124 (1977)
8. Internet Usage, http://www.internetworldstats.com/stats.htm
9. D. Johnson, A. Menezes, S. Vanstone, The elliptic curve digital signature algorithm (ECDSA). Int. J. Inf. Secur. **1**(1), 36–63 (2001)
10. L. Martin, M. Schertler, Using the Boneh-Franklin and Boneh-Boyen identity-based encryption algorithms with the cryptographic message syntax (CMS), in *IETF Request for Comments*, 5409, Jan 2009
11. National Bureau of Standards, Data Encryption Standard, FIPS PUB 46, Jan 1977
12. National Institute of Standards and Technology (NIST), FIPS PUB 46-3 – Data Encryption Standard (DES) (1999), http://csrc.nist.gov/publications/fips/fips46-3/fips46-3.pdf
13. National Institute of Standards and Technology (NIST), FIPS PUB 186-2 – Digital Signature Standard (DSS) (2000), http://csrc.nist.gov/publications/fips/archive/fips186-2/fips186-2.pdf
14. National Institute of Standards and Technology (NIST), FIPS PUB 197 – specification for the Advanced Encryption Standard (AES) (2001), http://csrc.nist.gov/publications/fips/fips197/fips-197.pdf
15. Privacy Breach, http://www.nytimes.com/2007/11/22/world/europe/22data.html
16. R.L. Rivest, M.J.B. Robshaw, R.Sidney, Y.L. Yin, The RC6 block cipher (1998), ftp://ftp.rsasecurity.com/pub/rsalabs/rc6/rc6v11.pdf
17. RSA Laboratories, PKCS #1 v2.1: RSA cryptography standard (2002), http://www.rsa.com/rsalabs/node.asp?id=2125
18. Serpent – A Candidate Block Cipher for the Advanced Encryption Standard, http://www.cl.cam.ac.uk/~rja14/serpent.html
19. P.W. Shor, Algorithms for quantum computation: discrete logarithms and factoring, in *Proceedings of the 35th IEEE Annual Symposium on Foundations of Computer Science*, Santa Fe, Nov 1994, pp. 124–134
20. The Clinton Administration's Policy on Critical Infrastructure Protection: Presidential Decision Directive 63 (1998), http://csrc.nist.gov/drivers/documents/paper598.pdf
21. The Legion of the Bouncy Castle, http://www.bouncycastle.org/
22. Twofish, http://www.schneier.com/twofish.html

Chapter 2
Certificates

A major task of a PKI is to provide authenticity proofs for public keys. Important tools that are used in such proofs are *certificates*. In this chapter we explain the concept of a certificate and the main certificate standards.

2.1 The Concept of a Certificate

Suppose that user Alice wishes to verify a digital signature issued by user Bob. She queries a directory service and obtains Bob's public key. As explained in Sect. 1.4, Alice needs to convince herself of the authenticity of that public key. This is what certificates are used for. Certificates are data structures that bind public keys to entities and that are signed by a third party. If Alice has a certificate for Bob's public key and if Alice trusts the third party that signed the certificate and also trusts the signature verification key of the third party, then verifying the signature of the certificate convinces Alice of the authenticity of Bob's public key. In this way, certificates reduce the trust in a public key of an entity to the trust in some authority. This reduction can be iterated. This is explained in Sect. 3.3 where certificate chains are introduced.

We list the minimum contents of a certificate.

1. The name of the *subject* to which the public key in the certificate is bound. This may also be a pseudonym.
2. The *public key* which is bound to the entity.
3. The *cryptographic algorithm* with which the public key is to be used.
4. The *serial number* of the certificate.
5. The *validity period* of the certificate.
6. The name of the *issuer* of the certificate that signed the certificate.
7. *Restrictions* that apply to the usage of the public key in the certificate. For example, the usage may be restricted to entity authentication.

J.A. Buchmann et al., *Introduction to Public Key Infrastructures*,
DOI 10.1007/978-3-642-40657-7_2, © Springer-Verlag Berlin Heidelberg 2013

As we will see in the next section, certificates may contain much more information. The certificate content is signed by the issuing third party and the signature is appended to the certificate.

Typically, the issuer of a certificate is different from the subject. However, there are *self-issued certificates* in which subject and issuer are the same. They are, for example, used if issuers change their policy. A special case of self-issued certificates are *self-signed certificates* which have the additional property that the certified public key is also the public key required to verify the signature on the certificate. Self-signed certificates are used to make public keys of an issuer available to an application that can only process public keys contained in certificates.

2.2 X.509 Certificates

The most important certificate standard is the X.509 standard, which is explained in this section.

X.509 certificates are specified in the X.509 recommendation [17] of the telecommunication standardization sector of the International Telecommunication Union (ITU-T). The profile of X.509 certificates that are used on the Internet is specified in [3].

In the ITU-T X.509 recommendation two types of certificates are defined. The first is that of a public key certificate that binds a public key to a subject as explained in Sect. 2.1. The second type is that of an *attribute certificate* that binds attributes to a subject. For example, an attribute certificate may bind the role of a system administrator to a subject. The subject may use this certificate to obtain certain privileges. Attribute certificates are discussed in Sect. 2.4.

In the literature, the notion of an X.509 certificate typically refers to a public key certificate as it is more important. In this book, we also adopt this convention. If nothing else is said, X.509 certificate always means X.509 public key certificate.

2.2.1 *Structure*

X.509 certificates are specified using the *abstract syntax notation version 1* (ASN.1) as a specification language. This language is widely used for the description of data structures. ASN.1 is standardized by the ITU-T in [15]. An ASN.1 tutorial can be found in [5]. With ASN.1 it is possible to describe complex data structures. Listing 2.1 shows the high-level ASN.1 specification of an X.509 certificate. A structural representation can be found in Fig. 2.1.

ASN.1 permits various encoding rules. X.509 certificates are encoded according to the *distinguished encoding rules* (DER) described in [16]. DER prescribe the unique encoding of any ASN.1 structure and are based on the *basic encoding rules* (BER).

```
Certificate ::=  SEQUENCE  {
  tbsCertificate        TBSCertificate,
  signatureAlgorithm    AlgorithmIdentifier,
  signatureValue        BIT STRING }
```

Listing 2.1 ASN.1 specification of an X.509 certificate

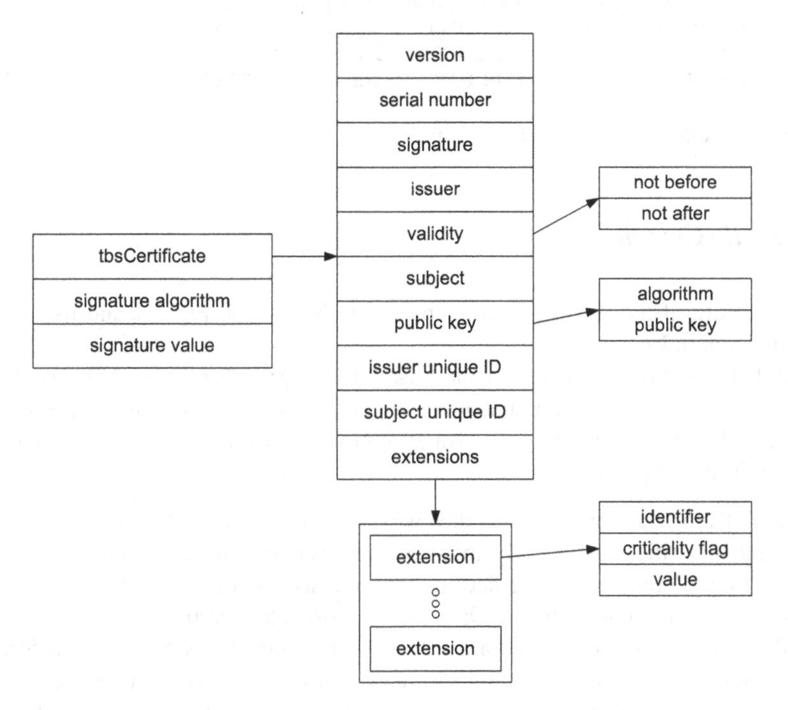

Fig. 2.1 The structure of an X.509v3 certificate

An X.509 certificate is an ASN.1 `SEQUENCE`. Such a sequence is an ordered list of elements. An X.509 certificate has three elements. The first element is `tbsCertificate` which is of type `TBSCertificate`. It contains the contents of the certificate as described in Sect. 2.1, for example, the public key and the name of its owner. The second element is `signatureAlgorithm` which describes the signature algorithm used by the third party to sign the certificate. This element contains the corresponding OID and the necessary parameters. It is of type `AlgorithmIdentifier`. The third element is `signatureValue`. It contains the signature of the certificate. It is of type `BIT STRING`.

In the following section we describe the certificate elements in more detail. They are also referred to as the *fields* of the certificate.

```
TBSCertificate  ::=  SEQUENCE  {
  version          [0]  EXPLICIT Version DEFAULT v1,
  serialNumber          CertificateSerialNumber,
  signature             AlgorithmIdentifier,
  issuer                Name,
  validity              Validity,
  subject               Name,
  subjectPublicKeyInfo SubjectPublicKeyInfo,
  issuerUniqueID   [1]  IMPLICIT UniqueIdentifier OPTIONAL,
  subjectUniqueID  [2]  IMPLICIT UniqueIdentifier OPTIONAL,
  extensions       [3]  EXPLICIT Extensions OPTIONAL }
```

Listing 2.2 ASN.1 specification of tbsCertificate

2.2.2 tbsCertificate

The structure of the tbsCertificate element can be seen in Fig. 2.1 and its ASN.1 specification in Listing 2.2.

All fields in this element except the `issuerUniqueID`, `subjectUniqueID`, and the `extensions` are mandatory, which means that they must be present in order for the certificate to be valid. An exception is the version field, which may be omitted if the version is 1.

version The `version` field specifies which version of the X.509 specification is used in this certificate. Currently, there are three versions of X.509 certificates. The first version X.509v1 was specified in 1988, the second version X.509v2 in 1993, and the third version X.509v3 in 1996. Today, most X.509 certificates use version 3. The differences of these versions are as follows. In contrast to X.509v1 and X.509v2 certificates, X.509v3 certificates may have extensions. Extensions of certificates are covered in Sect. 2.3. The difference between v1 and v2 X.509 certificates is that the latter are allowed to contain the issuerUniqueID and subjectUniqueID fields, which we discuss in a subsequent paragraph.

The value of the version field is an integer. If the certificate is v1 then this field can be omitted or the value of version is 0. For X.509v2 certificates the value of version is 1. For X.509v3 certificates this value is 2.

serialNumber The certificate issuer assigns a `serialNumber` to the certificate that is a non-negative integer encoded as a 20 bytes `ASN1Integer`. The serial number field is mandatory. The issuer must not assign the same serial number to more than one certificate. Therefore, the combination of the issuer name and the serial number uniquely identifies a certificate. This fact is used in many PKI protocols and applications. To meet this requirement, many issuers assign a starting number to their first certificate. Then the serial number is incremented by one each time a new certificate is issued. But other methods to assign serial numbers to certificates also exist.

Table 2.1 Typical attribute types

Attribute type	String representation	OID
countryName	C	2.5.4.6
organizationName	O	2.5.4.10
organizationalUnitName	OU	2.5.4.11
commonName	CN	2.5.4.3
localityName	L	2.5.4.7
stateOrProvinceName	ST	2.5.4.8

signature The issuer of an X.509 certificate signs the certificate. The mandatory field *signature* describes the signature algorithm that was used by the issuer to sign the certificate. The field is of type AlgorithmIdentifier. It contains the OID of the signature algorithm. OIDs for common signature schemes are listed in Table 1.5.

The signature field may also contain algorithm parameters that were used in the signature. This entry is optional. In most cases those parameters are not provided using this field, but are provided with the public key that is to be used to verify the certificate's signature. Providing parameter information may even be prohibited. For example, when ECDSA with SHA1 is used, the algorithm parameters must not be included (Sect. 2.2.3 of [14]).

Note that the signature algorithm specified in the signature field is also specified in the element signatureAlgorithm.

issuer The mandatory *issuer* field specifies the entity that issued the certificate and guarantees the correctness of its contents. The issuer is represented by an ASN.1 string called *distinguished name* (DN). Distinguished names are also used in other contexts. The structure of a DN is specified in [18]. An example of a DN is `CN=Alice, OU=Administration, O=TU Darmstadt, C=DE`. This DN describes a person with *common name* (CN) Alice, who belongs to the *organizational unit* (OU) "administration" of the *organization* (O) "TU Darmstadt" that operates in the *country* (C) Germany (DE as specified in ISO 3166 [9]). In our example, the DN reflects a logical hierarchy of a person belonging to an organizational unit which is part of an organization located in a country. In general, a DN is a sequence of attributes and their values. An attribute and its value are separated by a "=" sign stating that an attribute has a specific value. The attribute-value pairs are separated by commas. Table 2.1 shows typical attributes of a DN. Although DNs do not have to reflect a hierarchy, this has proved to be very useful.

Currently, the encoding of an X.509 DN must be `UTF8` or `PrintableString`. Other encoding such as `TeletexString` or `BMPString` may be used by older certificates. Therefore, client applications that use X.509 certificates should support all those encodings.

validity The *validity* field indicates the validity period of a certificate. This field contains the two dates `notBefore` and `notAfter`. The notBefore date is a point in time before which the certificate is not yet valid. The notAfter date is a point in time after which the certificate is not valid anymore. Between these two dates the

```
ECParameters ::= SEQUENCE {
  version    ECPVer,
  fieldID    FieldID,
  curve      Curve,
  base       ECPoint,
  order      INTEGER,
  cofactor   INTEGER OPTIONAL }
```

Listing 2.3 Parameters of an elliptic curve that may be included

certificate is valid unless it has been revoked. Revocation of certificates is discussed in Chap. 5. Until the year 2049 `UTCTime` encoding must be used for representing the two dates. From 2050, the `GeneralizedTime` encoding is to be used.

subject The *subject* field describes the owner of the certificate, that is, the entity that owns the private key corresponding to the public key contained in the certificate. This owner may also be described in the subject alternative name extension (see Sect. 2.3). If the owner is described only in the subject alternative name extension, then the subject field contains an empty sequence. Like issuers, subjects are described by distinguished names. If the subject of a certificate is the issuer of another certificate, then it is essential that the subject DN match the issuer DN of all certificates issued by this entity and that the subject DN not be empty. Examples of attributes that can be used to represent subject DNs are found in Table 2.1.

subjectPublicKeyInfo The `subjectPublicKeyInfo` field contains the public key that is certified by the certificate. The public key is described as a sequence containing the OID of an algorithm followed by optional parameters and the public key. An example of parameters for a public key used in elliptic curve cryptography can be found in Listing 2.3 (see also [14]). The public key is represented using the ASN.1 format. This format is given by its DER-encoding, which is a binary string.

issuerUniqueID and subjectUniqueID Although this is not recommended, it may happen that a distinguished name is assigned to different entities. For example, if a subject DN is used twice by an issuer, then the owner of the corresponding certificate is not uniquely determined by the subject DN. To make the owner description unique, the subjectUniqueID field may be added. The content of that field is a binary string that is a unique identifier for the owner of the certificate.

Likewise, several issuers may share the same DN. In this case the issuerUniqueID field resolves the situation. The subjectUniqueID and issuerUniqueID fields were introduced in the second version of X.509. Therefore, only X.509v2 and X.509v3 certificates may contain these fields. The use of non-unique distinguished names and of the subjectUniqueID and issuerUniqueID fields is not recommended because they make certificate use more complicated.

extensions The field extensions is discussed in Sect. 2.3.

```
Extension  ::=  SEQUENCE  {
  extnID      OBJECT IDENTIFIER,
  critical    BOOLEAN DEFAULT FALSE,
  extnValue   OCTET STRING }
```

Listing 2.4 Extension

2.2.3 *signatureAlgorithm*

As explained in the previous section, the signature algorithm that was used to sign the certificate is specified twice in an X.509 certificate, once in the tbsCertificate field and once in the signatureAlgorithm field. The reason for this remains unclear.

2.2.4 *signatureValue*

This field holds the signature on the tbsCertificate content of the certificate. It is represented as a bit string.

2.3 X.509 Certificate Extensions

When used in practice, the contents of X.509v1 and v2 certificates turned out to be insufficient. Therefore, X.509v3 certificates may contain extensions which support various PKI processes such as locating the issuer of a certificate. The ASN.1 structure of X.509 certificate extensions can be seen in Listing 2.4.

The first field in such an extension is `extnID`, which contains the OID of the extension. Next, any extension contains a criticality indicator *critical*. If its value is *true* then all applications that use this certificate must evaluate the extension. If an application is unable to do so, then it must consider the certificate to be invalid. The third field contains the DER-encoded ASN.1 structure of the extension description.

In the following we explain the 17 extensions that are defined in RFC 5280 [3]. In addition, any application may define and use its own extensions.

AuthorityKeyIdentifier The purpose of the `AuthorityKeyIdentifier` extension, also known as AKI extension or AKIE, is to support applications in identifying the public key of the issuer, which must be used to verify the certificate signature. The information in the issuer field may not be sufficient to identify this public key since an issuer may have several public keys. The authority key identifier extension must be present in any X.509v3 certificate unless the certificate is self-signed. Also, this extension must not be marked critical. For example, this extension is used by Microsoft [2]. The ASN.1 specification of the extension can

```
AuthorityKeyIdentifier ::= SEQUENCE {
  keyIdentifier             [0] KeyIdentifier           OPTIONAL,
  authorityCertIssuer       [1] GeneralNames            OPTIONAL,
  authorityCertSerialNumber [2] CertificateSerialNumber OPTIONAL
     }

KeyIdentifier ::= OCTET STRING
```

Listing 2.5 AKI extension

be seen in Listing 2.5. It supports two methods for identifying the issuer's public key. The first method assumes that an application has access to several public keys of the certificate issuer and must choose between them. This choice can be made by comparing the SHA-1 hash value stored in the `keyIdentifier` field with the SHA-1 hash values of the known public keys. Note that a successful comparison is not a proof of authenticity of the issuer public key. This authenticity must be established differently since it is the basis of the validation of the certificate. To save space, the contents of the keyIdentifier may be restricted to the 60 least significant bits of the hash value appended with 0100.

We describe the second method used by the authority key identifier extension to specify the verification key for the certificate signature. It describes this public key by presenting the distinguished name `authorityCertIssuer` of the issuer and the serial number `authorityCertSerialNumber` of a certificate that authenticates the public key. The authorityCertIssuer must coincide with the issuer of the certificate. Serial number and distinguished name uniquely determine such a certificate. From this certificate, an application can extract the relevant public key. All fields in this extension are optional. For example, only the distinguished name of the issuer may be present. However, this is not recommended since it may not identify the public key uniquely. This second specification method has a serious disadvantage. Suppose that an application attempts to verify a certificate using an authority key identifier extension that does not contain the hash value of a public key but points to an issuer public key certificate. Assume that the issuer public key certificate has expired. Then the certificate verification fails. However, the issuer public key may still be valid. For example, there may be a new certificate for that key. But due to the identification method, the verification process cannot find this new certificate. This disadvantage does not occur if a hash value is used to specify the issuer public key since hash values do not contain validity information.

Subject Key Identifier The `SubjectKeyIdentifier` extension of a certificate contains the hash value of the public key which is certified by the certificate. It is used by applications that compare the public key in the certificate to other public keys. The extension is very useful if the owner of the certificate has several public keys. As in the AuthorityKeyIdentifier extension, this hash value is either the full SHA-1 hash or the leading part of it. This extension must not be marked critical and

```
KeyUsage ::= BIT STRING {
  digitalSignature          (0),
  nonRepudiation            (1),
  keyEncipherment           (2),
  dataEncipherment          (3),
  keyAgreement              (4),
  keyCertSign               (5),
  cRLSign                   (6),
  encipherOnly              (7),
  decipherOnly              (8) }
```

Listing 2.6 Key usage extension

must appear in certificates whose public keys can be used to verify signatures on other certificates.

Key Usage The `KeyUsage` extension indicates what the public key contained in the certificate can be used for. Listing 2.6 shows the ASN.1 structure of this extension. The possible uses are the following.

digitalSignature The public key can be used to verify digital signatures, for example, in entity authentication protocols or for signing emails.

nonRepudiation The public key can be used to verify signatures that provide non-repudiation, as explained in Sect. 1.3.7.

keyEncipherment The public key may be used to encrypt symmetric session keys as explained in Sect. 1.3.5. It may also be used to encrypt private keys of asymmetric systems, for example in key delivery protocols.

dataEncipherment The public key may be used to encrypt data.

keyAgreement The public key may be used in a key agreement scheme.

keyCertSign The private key corresponding to the public key in the certificate may be used to sign certificates. The public key is then used to verify certificate signatures.

cRLSign The private key corresponding to the public key in the certificate may be used to sign certificate revocation lists. Such lists are discussed in Sect. 5.2.

encipherOnly This field is used when the keyAgreement usage option is selected.

decipherOnly This field is used when the keyAgreement usage option is selected.

Many clients and applications evaluate the key usage extension. A typical example is that of an email client which has access to several certificates of the same entity. The email client can tell by the key usage extension which certificate to use for encrypting emails or for verifying email signatures. Likewise, for certificate issuers the keyCertSign usage must be selected if the key usage extension is present.

The key usage extension must be consistent with the corresponding cryptographic algorithm and must not contain contradictions. For example, for a DSA key the key usage dataEncipherment is inappropriate since DSA is a signature algorithm. Also, encipherOnly and decipherOnly cannot be set at the same time.

Subject Alternative Name As explained in Sect. 2.2.2, the subject field contains the distinguished name of the public key owner. The `SubjectAlternativeName` extension binds additional names to the public key in the certificate. Typical names contained in this extension are email or IP addresses, domain names (DNS names) or uniform resource identifiers (URIs). For example, when a certificate is used for encrypting emails, the email address of the recipient may be contained in this extension. Also, if the public key in the certificate is used for entity authentication of the Web server of an organization, the DNS name or the IP address of that server is typically contained in this extension. Most clients that connect securely to such a server verify that the IP address or the DNS name of the server matches the IP address or DNS name contained in this extension.

Name Constraints This extension is used by the issuer of a certificate to provide limitations on the values of the subject field or subject alternative name extension in certificates issued using the key of this certificate. It may permit or exclude names. For example, a bank that issues certificates may be required by this extension to include the bank name in the certificates.

Issuer Alternative Name The structure and meaning of the `Issuer AlternativeName` extension is analogous to the subject alternative name extension. Note that this extension is not used in the certificate verification algorithm presented in Sect. 9.4.1.

Subject Directory Attributes This extension may contain attributes associated to the subject in the certificate. Examples are its initials or its nationality.

Extended Key Usage This extension associates the public key in the certificate with additional uses and functions that augment or replace the standard key usages specified in the key usage extension. Examples are server or client authentication, code signing, email security, time-stamping, and signing answers to certificate status queries.

Further Extensions Further extensions that will be discussed later are: the `CertificatePolicies` and `PolicyMappings` extensions in Sects. 8.2.1 and 8.2.2, respectively, the `BasicConstraints` extension in Sect. 3.3.1, the `PolicyConstraints` extension in Sect. 8.2.3, the `CRLDistributionPoints` extension in Sect. 5.3.1, the `Inhibit anyPolicy` extension in Sect. 8.2.4, and the extension `FreshestCRL` in Sect. 5.2.2.

The extensions discussed so far can be found in the X.509 specification [17] and are standard extensions of a certificate. Other extensions that may be defined for certain contexts or applications are called private extensions. In [3], where PKIs for the Internet are specified, two private extensions are described. These are the `AuthorityInformationAccess` and the `SubjectInformationAccess` extensions. They are discussed in Sects. 9.6.1 and 9.6.2, respectively.

2.4 Attribute Certificates

An *attribute certificate* is a digital document which is used by so-called attribute authorities (AA) to assign privileges to entities. For example, such certificates are used by mobile phone operating systems to verify that certain services have the right to be executed in the controlled environment of mobile phones. For example, if a third party wishes to access the address book of a mobile phone in order to include more information, an attribute certificate proves that this operation is permitted. In contrast to standard X.509 certificates, attribute certificates do not contain public keys. Instead, they bind an authorization to the owner of the certificate. Like standard X.509 certificates, attribute certificates are also signed by a trusted third party. Although such certificates are also specified in the X.509 standard, they are typically not called X.509 certificates. X.509 certificates are assumed to certify public keys.

Attribute certificates are used in conjunction with X.509 certificates. For example, consider a firmware update for a mobile phone. It is signed by its issuer and the signature verification key is authenticated by a certificate. In addition, an attribute certificate may specify whether or not this update may be used for a certain type of mobile phone.

There are several reasons for separating public key certificates from attribute certificates. The authorization may have a much shorter lifetime than the public key in the X.509 certificate. For example, firmware updates may be changed frequently. The authorization to install the update expires when a new update is published. However, the public key of the update publisher may be valid much longer. Also, attribute certificates may replace complicated access lists. In the case of firmware updates, the mobile device can just evaluate the attribute certificate instead of checking by itself whether the update is appropriate.

The profile of X.509 attribute certificates is specified in [17] and [8]. The structure of an attribute certificate resembles that of an X.509 certificate. It is a sequence: some content that is signed, the signing algorithm, and the signature. It has basic fields: the owner, the issuer, the serial number, the validity period, and the privileges (authorization information) of the owner. It may also contain extensions. The ASN.1 specification of an X.509 attribute certificate can be seen in Listing 2.7. An infrastructure that is based on attribute certificates is called privilege management infrastructure (PMI).

2.5 CV Certificates

Card verifiable (CV) certificates [4, 10] are very compact public key certificates. CV certificates avoid the overhead which comes with flexible general-purpose certificate formats such as X.509. Their purpose is to be verifiable by resource-restricted devices such as smart cards, in particular to enable the authentication of an entity

```
AttributeCertificate ::= SEQUENCE {
  acinfo                  AttributeCertificateInfo,
  signatureAlgorithm      AlgorithmIdentifier,
  signatureValue          BIT STRING }

AttributeCertificateInfo ::= SEQUENCE {
  version                 AttCertVersion -- version is v2,
  holder                  Holder,
  issuer                  AttCertIssuer,
  signature               AlgorithmIdentifier,
  serialNumber            CertificateSerialNumber,
  attrCertValidityPeriod    AttCertValidityPeriod,
  attributes              SEQUENCE OF Attribute,
  issuerUniqueID          UniqueIdentifier OPTIONAL,
  extensions              Extensions OPTIONAL }
```

Listing 2.7 Attribute certificate

```
cvcBody ::=  SEQUENCE {
   profileId         UNSIGNED INTEGER,
   issuer            CHARACTER STRING,
   pubKey            SEQUENCE,
   subject           CHARACTER STRING,
   chat              SEQUENCE,
   notBefore         DATE,
   notAfter          DATE }
```

Listing 2.8 ASN.1 specification of a CV certificate

to the device. After a successful authentication, the entity may use certain access-protected functions of the device. CV certificates only contain the most important fields, which are:

Certificate authority A reference to the certificate issuer.
Certificate holder public key The certified public key.
Certificate holder A reference to the certificate subject.
Certificate holder authorization The subject's access rights to the device.
Certificate effective date Start of the validity of the certificate.
Certificate expiration date End of the validity of the certificate.

The corresponding ASN.1 structure of this *self-descriptive variant* can be found in Listing 2.8.

To make CV certificates even more compact, the *not self-descriptive variant* of CV certificates can be used. It integrates parts of the certificate into the signature by using signatures with message recovery as specified in [11]. In this variant the CV certificate only consists of a sequence of strings. The corresponding format must be made known to the evaluating device beforehand.

2.6 PGP Certificates

Pretty good privacy (PGP), which was invented 1991 by Phil Zimmermann, establishes trust in the ownership of public keys using a social network. Trust in the ownership of a public key is established by trustworthy participants of the network certifying this ownership. This is what PGP certificates are used for. PGP is discussed in detail in Sect. 3.2.

PGP certificates are specified in the OpenPGP specification [1]. They contain at least one public key, which is described in a so-called *public key packet*. A public key packet includes:

- The PGP version number that indicates which version of PGP was used to create the key associated with the certificate.
- The creation time of the key.
- The public key together with the algorithm in which the key is to be used.

A PGP certificate includes:

- A public key packet.
- Optionally: Signatures to invalidate this certificate.
- Identity information about the certificate holder, for example, name and email address.
- A digital signature of the public key packet and the identity information. This signature is issued using the private key that corresponds to the public key in the certificate.
- Optionally: signatures of the public key packet and the identity information of other PGP users. These signatures certify the authenticity of the public key contained in the certificate. They are required to establish trust in the PGP web of trust as explained in Sect. 3.2.
- Optionally: other identity information. This information may also be signed.
- Optionally: attributes of the user. They may also be signed.
- Optionally: other public keys of the user. They may also be signed.

The current version of PGP certificates is version 4. PGP version 3 certificates that may only contain one public key must not be generated anymore. However, applications should be able to process them. The structure of PGP version 4 certificates is displayed in Fig. 2.2.

2.7 Other Certificates

There are several other certificate types that are used in certain contexts or applications. We briefly discuss these certificates.

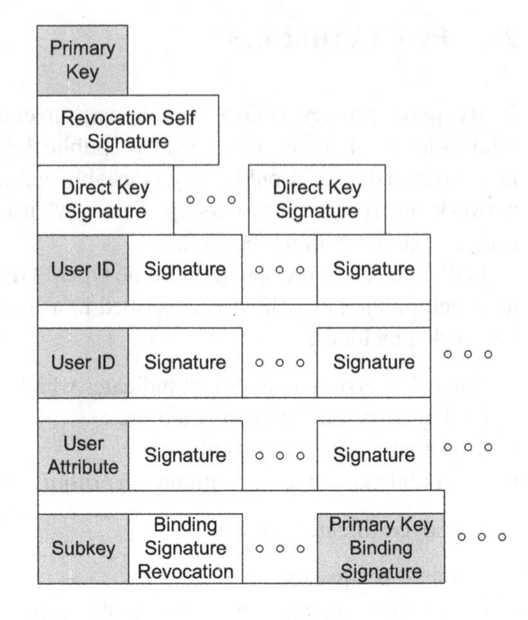

Fig. 2.2 A PGP V4 certificate; the shaded components are mandatory

2.7.1 WAP Certificates

The wireless application protocol (WAP) is a specification for developing applications in wireless communication networks. Such applications have to satisfy special requirements that account for the properties of devices used in such networks, such as limited storage capabilities. The corresponding WAP certificates (see [20]) are modified X.509 certificates with reduced storage requirements. For example, in WAP certificates serial numbers of user certificates must not be longer than 8 bytes. To provide extra information that does not fit into the WAP certificate, such certificates may have a `domainInformation` extension which contains a pointer to extensions that are not contained in the certificate but are associated to it.

2.7.2 SPKI Certificates

An alternative to the X.509 standard is simple public key infrastructure (SPKI) [6,7]. It was specified by the corresponding working group at the IETF, which no longer exists. SPKI uses the so-called SPKI certificates. Their main purpose is to support authorization.

There are two types of SPKI certificates, the authorization certificate and the name certificate. SPKI authorization certificates have issuer and subject fields. The issuer is represented by a public key or its hash value. The subject may contain the hash of an object (for example, a software package), a name, a public key (or its hash value), information about the key holder intended for human readability, and data that support multi-party computation. Such certificates also have fields that specify their version and their validity period and that provide human readable content and additional information about the issuer (other certificates that authorize this issuer to issue certificates) and the subject (for example, its public key). Authorization certificates also support delegation.

Name certificates resemble authorization certificates but do not contain authorization or delegation statements or additional information about the issuer or the subject. They just bind names to public keys. Since issuers can use names of any form, such certificates are easier to use for humans. One possible encoding of SPKI certificates is XML [12].

For further reading on SPKI we refer the reader to [6, 7], and [19].

2.7.3 Traceable Anonymous Certificate

When X.509 certificates are used by applications, the user's identity can be easily obtained from the certificate. In certain applications users may want to remain anonymous. On the other hand, under certain conditions an application may have the right to learn the user's real identity. This is supported by *traceable anonymous certificates* as defined in [13]. Such certificates use the X.509 format. However, the collaboration of two issuers guarantees the desired anonymity properties. Those issuers are the *blind issuer* and the *anonymity issuer*. The user and the two authorities interact in a protocol in which the user creates her or his key pair, the blind issuer verifies the real identity of the user, and the anonymity issuer issues the certificate in which the subject is a pseudonym. The blind issuer and the anonymity issuer can only link the user to the certificate if they collaborate.

2.8 Exercises

2.1. How many versions of X.509 certificates exist? What are the differences between them?

2.2. In the three certificates shown in Fig. 2.3 the values of the keyIdentifier (keyId), authorityCertIssuer (aci), and authorityCertSerialNumber (acsn) of the authority key identifier extension are missing. Enter appropriate values.

2.3. Answer the following questions.

1. The authority key identifier extension and subject key identifier extension both have a field called keyIdentifier. Can these fields have the same content in the same certificate?
2. Can two different certificates have the same keyIdentifier in the authority key identifier extension?

Certificate 1	
Serial No.:	26540
Issuer:	CN=Test CA
NotBefore:	2004-04-03
NotAfter:	2005-04-03
Subject:	CN=Alice
Public Key:	key-0x4D367AB9
X509v3Extensions:	
KeyUsage: critical digitalSignature, dataEncipherment	
Subject Key Identifier: keyId: 12:AB:45:76:F8:98	
Authority Key Identifier: keyId: aci: acsn:	

Certificate 3	
Serial No.:	34
Issuer:	CN=Master CA
NotBefore:	2003-11-15
NotAfter:	2008-11-10
Subject:	CN=Test CA
Public Key:	key-0x347893B2
X509v3Extensions:	
KeyUsage: critical keyCertSign, cRLSign	
Subject Key Identifier: keyId: BE:76:34:4E:60:34	
Authority Key Identifier: keyId: aci: acsn:	

Certificate 2	
Serial No.:	1
Issuer:	CN=Master CA
NotBefore:	2003-11-14
NotAfter:	2008-11-14
Subject:	CN=Master CA
Public Key:	key-0x18FF6542
X509v3Extensions:	
KeyUsage: critical keyCertSign, cRLSign	
Subject Key Identifier: keyId: 11:23:34:AB:65:F0	
Authority Key Identifier: keyId: aci: acsn:	

Fig. 2.3 Certificates with missing AKI values

Table 2.2 Key usage extension values

(0)								(8)
1	1	0	1	0	0	0	0	0

3. Can two different certificates have the same keyIdentifier in the subject key identifier extension?

2.4. In Table 2.2, values of the key usage extension are given. For what purposes can the key in the corresponding certificate be used? For which algorithm is this possible?

2.5. The certificates in Fig. 2.4 have the same serial number. Is this permitted?

Certificate 1	
Serial No.:	2
Issuer:	E
NotBefore:	2003-02-01
NotAfter:	2005-12-31
Subject:	I
Public Key:	key-0x384756AB

Certificate 2	
Serial No.:	2
Issuer:	B
NotBefore:	2003-02-01
NotAfter:	2005-12-31
Subject:	F
Public Key:	key-0x4569DEFA

Certificate 3	
Serial No.:	2
Issuer:	G
NotBefore:	2003-02-01
NotAfter:	2005-12-31
Subject:	L
Public Key:	key-0x8269AEB8

Fig. 2.4 Certificates with same serial numbers

References

1. J. Callas, L. Donnerhacke, H. Finney, D. Shaw, R. Thayer, OpenPGP message format, in *IETF Request for Comments*, 4880, Nov 2007
2. Certificate Status and Revocation Checking, http://social.technet.microsoft.com/wiki/contents/articles/certificate-status-and-revocation-checking.aspx
3. D. Cooper, S. Santesson, S. Farrell, S. Boeyen, R. Housley, W. Polk, Internet X.509 public key infrastructure certificate and certificate revocation list (CRL) profile, in *IETF Request for Comments*, 5280, May 2008
4. Comité Européen de Normalisation (CEN), Application interface for smart cards used as secure signature creation devices—part 1: basic requirements. CEN Workshop Agreement (2004), ftp://ftp.cenorm.be/PUBLIC/CWAs/e-Europe/eSign/cwa14890-01-2004-Mar.pdf
5. O. Dubuisson, *ASN.1—Communication Between Heterogeneous Systems* (Morgan Kaufmann, San Diego, 2000)
6. C. Ellison, SPKI requirements, in *IETF Request for Comments*, 2692, Sept 1999
7. C. Ellison, B. Frantz, B. Lampson, R. Rivest, B. Thomas, T. Ylonen, SPKI certificate theory, in *IETF Request for Comments*, 2693, Sept 1999
8. S. Farrell, R. Housley, An internet attribute certificate profile for authorization, in *IETF Request for Comments*, 3281, Apr 2002
9. International Organization for Standardization (ISO), English country names and code elements, http://www.iso.org/iso/english_country_names_and_code_elements
10. International Organization for Standardization ISO, ISO/IEC 7816-8: identification cards – integrated circuit(s) cards with contacts – part 8: security related interindustry commands. International Standard, Oct 1999
11. International Organization for Standardization ISO, ISO/IEC 9796: information technology – security techniques – digital signature schemes giving message recovery – parts 1–3. International Standard, 1999–2002
12. J. Paajarvi, XML encoding of SPKI certificates (2000), http://xml.coverpages.org/draft-paajarvi-xml-spki-cert-00.txt
13. S. Park, H. Park, Y. Won, J. Lee, S. Kent, Traceable anonymous certificate, in *IETF Request for Comments*, 5636, Aug 2009
14. W. Polk, R. Housley, L. Bassham, Algorithms and identifiers for the internet X.509 public key infrastructure certificate and certificate revocation list (CRL) profile, in *IETF Request for Comments*, 3279, Apr 2002

15. Recommendation X.680 ITU-T, Information technology – abstract syntax notation one (ASN.1): specification of basic notation, July 2002
16. Recommendation X.690 ITU-T, Information technology – ASN.1 encoding rules: specification of basic encoding rules (BER), canonical encoding rules (CER) and distinguished encoding rules (DER), July 2002
17. Recommendation X.509 ITU-T, Information technology – open systems interconnection – the directory: public-key and attribute certificate frameworks, Aug 2005
18. Recommendation X.501 ITU-T, Information technology – open systems interconnection – the directory: models, Nov 2008
19. SPKI/SDSI Certificate Documentation, http://world.std.com/~cme/html/spki.html
20. WAP Forum, WAP certificate and CRL profiles specification, http://www.openmobilealliance.org/tech/affiliates/wap/wap-211-wapcert-20010522-a.pdf

Chapter 3
Trust Models

Public key cryptography can only be used in practice if users trust the authenticity of public keys. In this chapter we explain models and infrastructures that allow us to establish trust in the authenticity of public keys.

3.1 Direct Trust

Direct trust is the most basic trust model. In fact, it is required by all other trust models to initialize trust. We start by explaining this model in an example. Like other operating systems, many Linux variants allow the installation of additional software such as updates or services from servers located on the Internet. The authenticity of those software packages is established by a digital signature. The verification of the signature requires a public key. Frequently, this key is provided on the original Linux distribution CD or DVD, typically stored in PGP format. The authenticity of this key is guaranteed by the authenticity of the CD or DVD and by the difficulty of overwriting it. In such a situation, the trust in the public key is called *direct* since it is directly obtained from its owner (the Linux distributor). No third party is involved in authenticating the key.

More generally, we say that trust in the authenticity of a public key is direct if the public key is directly obtained from the key owner or its owner directly confirms the authenticity of the key in a way that is convincing for the user.

We present an example of a public key which is obtained from another source and which is authenticated by the owner. The email client Thunderbird comes with many pre-installed certificates that are trusted by default. This is shown in Fig. 3.1.

Those public keys are used to verify signatures on certificates which, in turn, authenticate public keys of end users or intermediate entities (cf. Sect. 3.3). More than 150 public keys are already contained in Thunderbird upon installation. They are stored in self-signed certificates in the Thunderbird key repository. Their authenticity can be verified as follows. Thunderbird computes a fingerprint of the

J.A. Buchmann et al., *Introduction to Public Key Infrastructures*,
DOI 10.1007/978-3-642-40657-7_3, © Springer-Verlag Berlin Heidelberg 2013

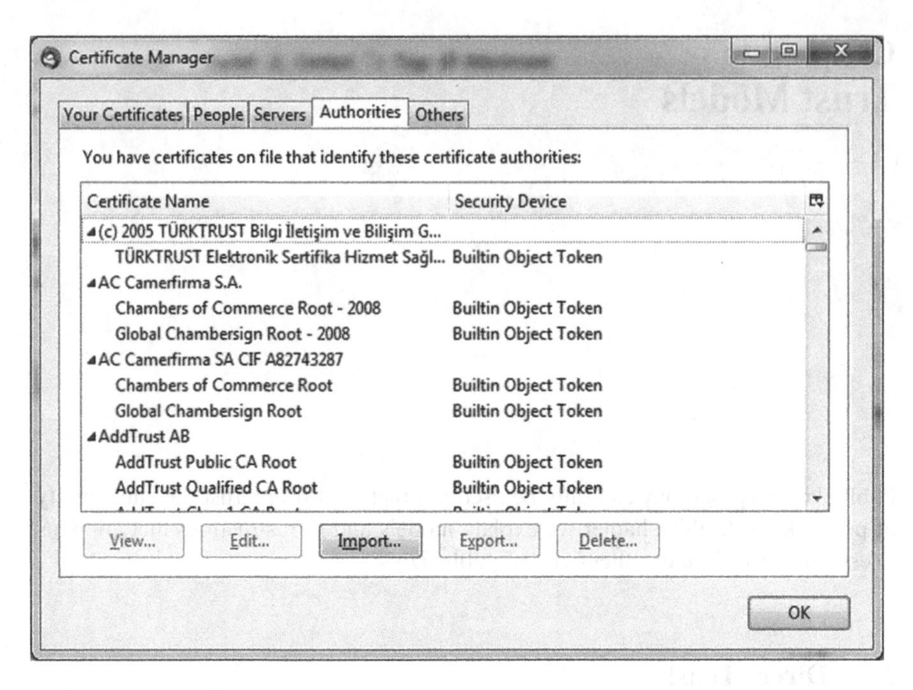

Fig. 3.1 Thunderbird's certificate manager

self-signed certificate that contains the public key. Figure 3.2 shows the so-called *German national Root-CA* certificate for legally binding electronic signatures as presented by Thunderbird's certificate viewer. The last section contains two different fingerprints. The first one is computed using the SHA1 hash algorithm and the second one is computed using the MD5 hash algorithm, which is no longer considered to be secure. One of these fingerprints is compared with the certificate fingerprint which can be obtained directly from the party that issued the certificate, for example from a customer hotline or a Web page. The fingerprint may also be printed on a business card or published in a newspaper or an official government publication. Figure 3.3 shows the Web page of CAcert [1], where the fingerprints of important CAcert certificates are published. CAcert is a community-driven service providing free certificates.

This procedure establishes trust in public keys because an additional communication channel is used to check the correctness of the fingerprint. Controlling two communication channels is considered to be much harder for an adversary than one.

Many applications use pre-installed public keys. For example, operating systems, email clients, and Web browsers. Most users trust those keys without checking them. However, as explained above, any user can verify the fingerprints of the corresponding certificates. This is explicitly encouraged by certain applications. For example, when a self-signed certificate is installed in a Windows operating system a dialog window appears that notifies the user about the certificate's fingerprint.

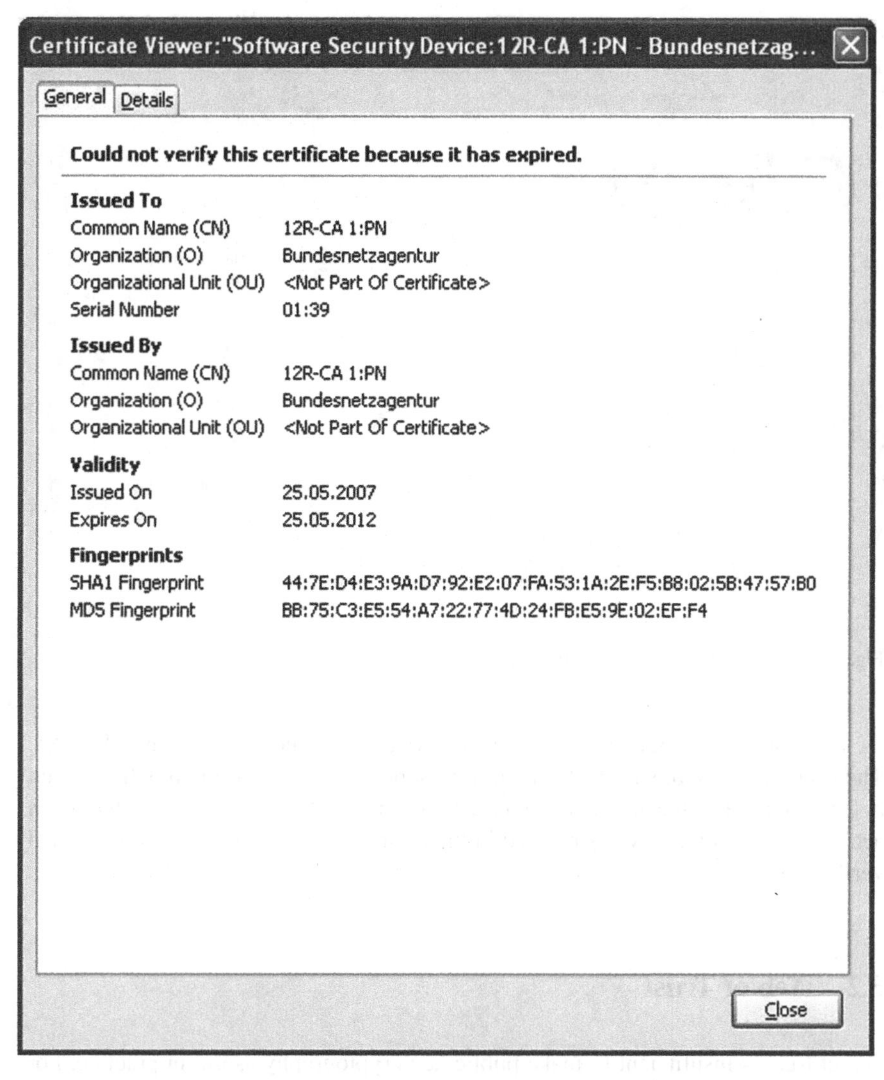

Fig. 3.2 German national Root-CA certificate in Thunderbird

Figure 3.4 shows this dialog as presented by Windows 7. It recommends checking the correctness of the fingerprint by contacting the owner of the public key.

Another example of the use of direct trust is the secure shell (SSH) [8] application. It is a popular program for connecting securely to remote servers. After a successful authentication, users are able to access resources of the remote system. Typically, user authentication is based on user name and password or on a public key mechanism. Servers are always authenticated by a public key mechanism. When a user attempts a login to the remote server for the first time, this server displays

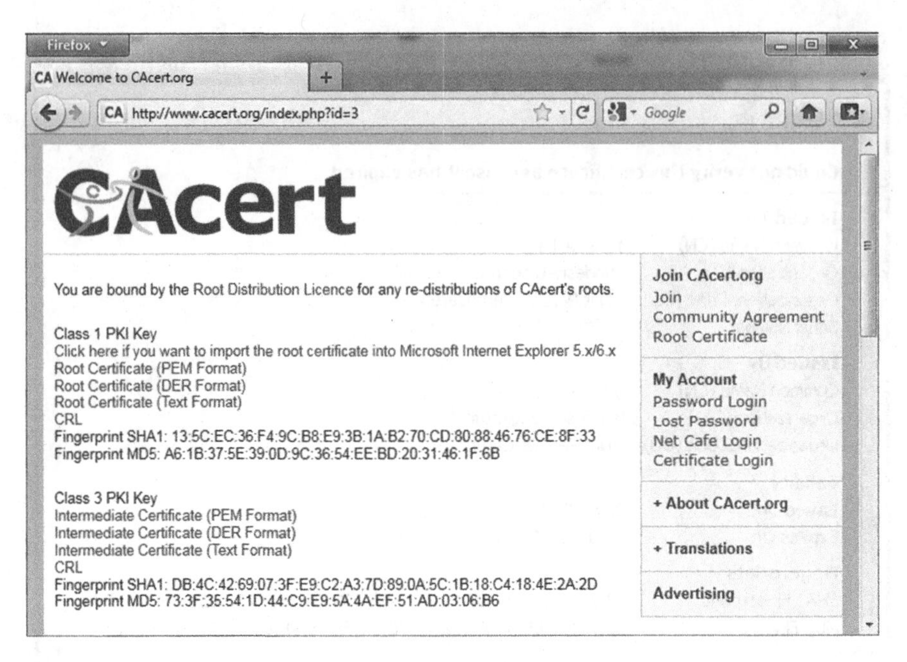

Fig. 3.3 Fingerprints of CAcert certificates

the fingerprint of its public key. The respective dialog window is shown in Fig. 3.5. The users may establish direct trust in this public key by verifying this fingerprint before continuing the login procedure. If the public key of the server has been accepted once, it will always be used in the login procedure without any additional verification.

3.2 Web of Trust

Direct trust is insufficient to make public key cryptography useful in practice. For example, if a user wants to send an encrypted email to another user, establishing direct trust requires the sender to contact the receiver prior to sending the email in order to obtain the receiver's public encryption key. Although this may be possible, it slows down email communication considerably. Therefore, other trust models have been introduced.

One of these alternative trust models is the web of trust. It was invented in 1992 by Phil Zimmermann in his system PGP [7]. The web of trust model is inspired by social trust networks. In the web of trust, users trust a public key if they have obtained it from its owner directly (direct trust) or if a sufficient number of trustworthy users recommend trusting the key.

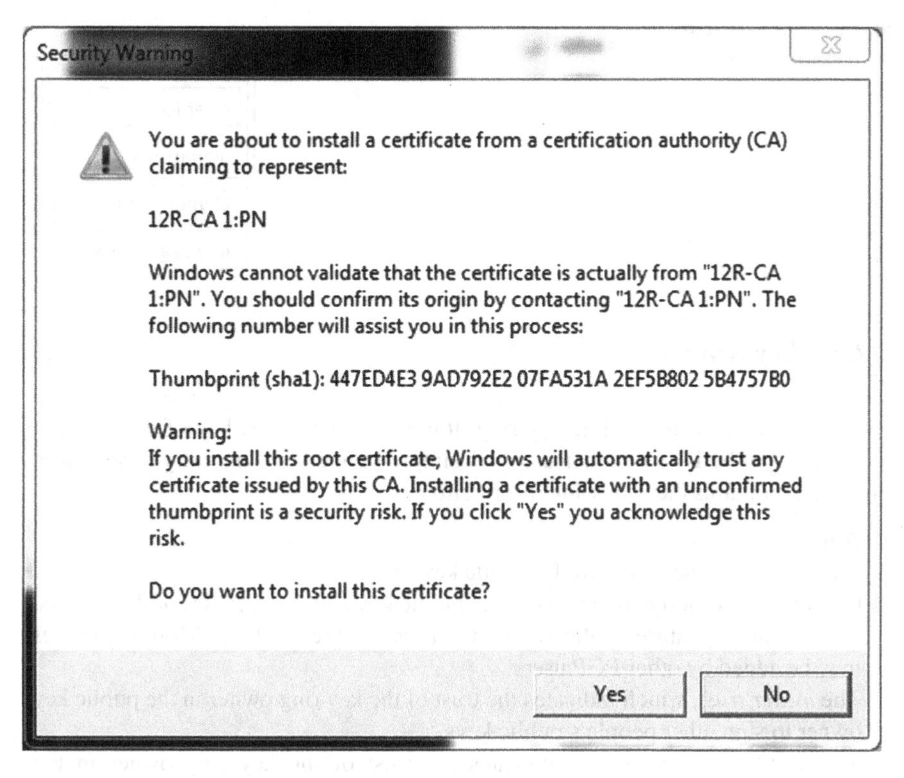

Fig. 3.4 Fingerprint dialog in Windows 7

```
$ ssh userver1.cdc.informatik.tu-darmstadt.de
The authenticity of host 'userver1.cdc.informatik.tu-darmstadt.de (130.83.167.15)' can't be
established.
RSA key fingerprint is b2:e8:ac:0c:f7:bc:46:73:78:af:08:cf:54:b4:6e:17.
Are you sure you want to continue connecting (yes/no)?
```

Fig. 3.5 Fingerprint dialog in SSH

In the following sections we explain in detail how the web of trust is implemented in PGP. The basic idea of PGP is that users who trust in a public key sign this public key. Such signatures are called *validity signatures*. PGP participants trust in a public key if it is signed by sufficiently many users whom they trust to sign keys of other users. Our explanation follows RFC 4880 [2], which defines the OpenPGP standard for the Internet. The corresponding open-source implementation is *GNU privacy guard* (GPG).

Fig. 3.6 A simple PGP public key ring entry

3.2.1 Key Ring

Every PGP user owns a public key ring. It contains the public key of the key ring owner and public keys of other users. A simple public key ring entry is shown in Fig. 3.6. It contains the following information.

- A public key.
- The user ID of the owner of the public key.
- One or more validity signatures of the public key and the user IDs of the signers. At least the signature of the key owner must be present here. More signatures may be added by other PGP users.
- The *owner trust*, which indicates the trust of the key ring owner in the public key owner to sign other people's public keys.
- The *key legitimacy*, which indicates the trust of the key ring owner in the authenticity of the public key in this entry. It is also called *key validity*.

The implementation details of the public key ring are not specified. However, there is a standardized way to export and import public keys using *Transferable Public Keys*, which are simply concatenated and stored in a single file. Export and import of private keys is done analogously using *Transferable Secret Keys*. Transferable public and secret keys can even be combined in one file.

The contents of the first two entries of a PGP key ring do not need further explanation. So the next sections only discuss the last three fields of PGP key rings.

Signatures Upon generating their key pairs, users sign their own public keys. The owner's signature is always attached to the public key and is present in any key ring that contains the public key. It is called *self-signature*. Other users may add their own signatures to certify the authenticity of public keys. Before they do so they are expected to establish direct trust in the signed public keys. For example, in the PGP community, key signing parties are organized in which users sign each other's keys.

Owner Trust The owner trust of a public key indicates how much trust the key ring owner has in the public key owner to sign public keys of other users. Therefore it is assigned by the key ring owner to each public key in his or her key ring. It can take the following five values.

- *Ultimate* owner trust is assigned to key ring owners' own public keys since they have unlimited trust in themselves.
- The owner trust is set to *complete* if the key ring owner trusts fully the public key owner to sign other public keys. This means that the key ring owner assumes that the public key owner very carefully checks the identity of the owners of other public keys before signing them.
- The owner trust is set to *marginal* if the key ring owner only trusts marginally the public key owner to sign other public keys. This means that the key ring owner assumes that the public key owner checks the identity of the owners of other public keys carelessly before signing them.
- If the key ring owner does not trust the public key owner to sign other public keys, the owner trust is set to *none*.
- If the key ring owner has no information how the public key owner checks the identity of the owners of other public keys before signing them, the owner trust is set to *unknown*.

Key Legitimacy The key legitimacy of a public key determines the trust of the owner of the public key ring in the authenticity of this public key, that is, in the fact that this public key belongs to the entity described in the user ID field of the corresponding entry in the key ring. The key legitimacy can take three values.

- The value *complete* indicates that the key ring owner is convinced that the public key belongs to the user described in the user ID field of the public key.
- The key legitimacy is *marginal* if the key ring owner is only marginally convinced that the public key belongs to the user described in the user ID field of the public key.
- The key legitimacy is set to *none* if the key ring owner does not know whether the public key belongs to the user described in the user ID field.

While owner trust is chosen by the key ring owners, key legitimacy is calculated when a public key is imported into the key ring. This calculation works as follows. Suppose that a key K is imported into the key ring of user Alice. This key comes with a number of signatures, one by the owner of key K and potentially several additional signatures by other people. We explain how the key legitimacy of K is determined.

The key legitimacy of K is set to complete if the key K is signed by Alice. In this case, Alice has verified the authenticity of K herself. She trusts this key directly.

The key legitimacy of K is also set to complete if there is at least one signature of K such that the corresponding verification key K' satisfies the following three conditions.

- K' is in Alice's public key ring.
- The key legitimacy of K' in Alice's key ring is complete.
- The owner trust of the owner of K' in Alice's key ring is complete.

This rule reflects the situation in which a public key is signed by a person who is completely trusted by Alice to sign other people's public keys.

Fig. 3.7 Alice's public key ring

	Public key ring of **Alice**		
1	Public key Owner		: Alice
	Owner Trust / Key Legitimacy		: ultimate / complete
	1	Signer / Trust in Signer	: Alice / ultimate
2	Public key Owner		: Bob
	Owner Trust / Key Legitimacy		: complete / complete
	1	Signer / Trust in Signer	: Bob / complete
	2	Signer / Trust in Signer	: Alice / ultimate
3	Public key Owner		: Carl
	Owner Trust / Key Legitimacy		: marginal / complete
	1	Signer / Trust in Signer	: Carl / marginal
	2	Signer / Trust in Signer	: Bob / complete
4	Public key Owner		: Diana
	Owner Trust / Key Legitimacy		: marginal / marginal
	1	Signer / Trust in Signer	: Diana / marginal
	2	Signer / Trust in Signer	: Carl / marginal
	3	Signer / Trust in Signer	: Oscar / –
5	Public key Owner		: Paul
	Owner Trust / Key Legitimacy		: none / none
	1	Signer / Trust in Signer	: Diana / marginal
	2	Signer / Trust in Signer	: Oscar / –

A third way for the key legitimacy of K to be set to complete is when there are at least two signatures of K such that the corresponding verification keys K_1 and K_2 satisfy the following three conditions:

- K_1 and K_2 are in Alice's key ring.
- The key legitimacy of K_1 and K_2 is complete.
- The owner trust of the owners of K_1 and K_2 is marginal.

The idea of this rule is that two signatures of users who are marginally trusted to sign other people's public keys is just as good as the signature of one user who is completely trusted. There are variants of this rule where more than two marginally trusted signatures are required. However, the rule presented here is mostly used.

The key legitimacy of K is set to marginal if none of the above rules holds and there is at least one signature of K such that the corresponding verification key K' satisfies the following three conditions.

- K' is in Alice's public key ring.
- The key legitimacy of K' in this key ring is complete.
- The owner trust of the owner of K' in the key ring is marginal.

In all other situations, the key legitimacy of K is set to none.

Example 3.1. Consider the public key ring of Alice as shown in Fig. 3.7. This key ring contains the keys of Alice, Bob, Carl, Diana, and Paul. Alice assigns owner trusts complete to Bob, marginal to Carl, marginal to Diana, and none to Paul. Alice has signed Bob's public key. Since the owner trust of Alice in herself is ultimate, the key legitimacy of Bob's key is set to complete. Bob has signed Carl's key. Alice's owner trust in Bob is complete. Therefore, the key legitimacy of Carl's public key is set to complete. Next, Diana's public key is signed by Carl and Oscar. Oscar's public key is not in Alice's public key ring. Carl's public key is in Alice's public key ring, its key legitimacy is complete and its owner trust is marginal. Hence, the key legitimacy of Diana's public key is set to marginal. Paul's public key is signed by Diana and Oscar. Both keys do not contribute to the key legitimacy of Paul's public key, since Oscar's public key is not in Alice's public key ring and the key legitimacy of Diana's key is marginal. Therefore, the key legitimacy of Paul's public key is set to none.

In order for secure communication to become possible with PGP, public keys, including their signatures, must be available to all users. In PGP this is achieved by users mutually exchanging their public keys or by publishing public keys on *key servers*. A key server is a publicly available directory service that allows everyone to upload and download public keys, including their signatures. Examples of publicly available PGP key servers are pgpkeys.pca.dfn.de, pool.skskeyservers.net, subkeys.pgp.net, and pgp.mit.edu.

3.2.2 Trust Signatures

We have seen that the web of trust lets users choose the trust in public key owners to sign public keys of others. This trust may also be computed using so-called *trust signatures*. A trust signature of a public key not only asserts that the key belongs to its claimed owner, but it also asserts that this owner is trustworthy to sign other keys. More precisely, a trust signature of a public key assigns two values to this key (and thereby to its owner): the *trust amount* and the *trust level*. Both are integers between 0 and 255. The trust amount corresponds to the owner trust. Usually, a trust amount less than 120 indicates partial owner trust; higher values indicate full owner trust. The meaning of the trust level is the following. Suppose that Alice has generated a trust signature of Bob's key with trust amount 200.

- If the trust level is 0, then the trust signature asserts that Alice has convinced herself of the validity of Bob's key and has owner trust 200 in Bob. This has the same meaning as an ordinary validity signature.
- If the trust level is 1, then the trust signature asserts that Alice has convinced herself of the validity of Bob's key and has owner trust 200 in Bob. Also, Alice imports Bob's trust amount values for all keys for which Bob has issued level 0 trust signatures. In this case, Bob is called a *trusted introducer* for Alice.
- Generally, trust level n means that Alice imports Bob's trust amount values for all public keys for which Bob has issued trust signatures with trust level less than n. For trust levels greater than 1, Bob is called a *meta introducer* for Alice.

Example 3.2. Figure 3.8 shows a trust chain with a meta introducer and a trusted introducer. Alice signs Bob's key with trust level 2 and trust amount 200, which makes Bob a meta introducer for Alice. Bob signs Carl's key with level 1 and trust

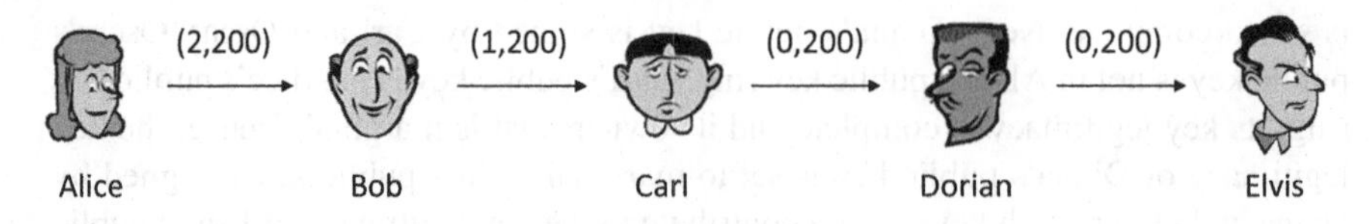

Fig. 3.8 A trust chain with a meta introducer and a trusted introducer

200, which makes Carl a trusted introducer for Bob. As Bob is a meta introducer for Alice, Carl is also a trusted introducer for Alice. Carl signs Dorian's key with level 0 and trust 200. As Carl is a trusted introducer for Alice, she accepts Dorian's key as valid. Note that Alice has never assigned any owner trust value to Carl explicitly. Nevertheless, she relies on his signature. She trusts Carl because he is a trusted introducer of Bob. But, as Alice decided to trust Bob with level 2 only, the trust chain ends here. In order to accept Elvis' key as valid, Alice has to manually set her owner trust in Dorian to complete.

3.2.3 *Probabilistic Trust Model for GnuPG*

The GNU project implements OpenPGP in GnuPG (GPG) [4]. In standard GPG, key legitimacy is calculated as a function of a finite set of values for the owner trust. In [5] the authors propose a probabilistic model for determining key legitimacy that is based on network reliability techniques.

3.3 Hierarchical Trust

The web of trust is quite convenient. It easily scales and requires no centralized infrastructure. But the applicability of the web of trust is limited. Signers of public keys typically do not accept (legal) liability for the authenticity of the public keys which they sign. Therefore, the web of trust appears to be inadequate in a business context. For example, consider a user who verifies the authenticity of a home banking Web page before entering a secret password. He does this by verifying a digital signature of the Web page. In such a context it would be preferable that the public verification key for the signature of the Web page is certified by some authority that can be made liable if the key turns out not to be authentic. In the hierarchical trust model, certificate signers accept such liability. Also, in a hierarchical PKI, trust in public keys depends on the trust in a uniquely determined certificate signer, the so-called *trust anchor*.

In a hierarchical PKI, public keys are certified by *certification authorities* (CAs). CAs implement a process for verifying the authenticity of public keys, which they

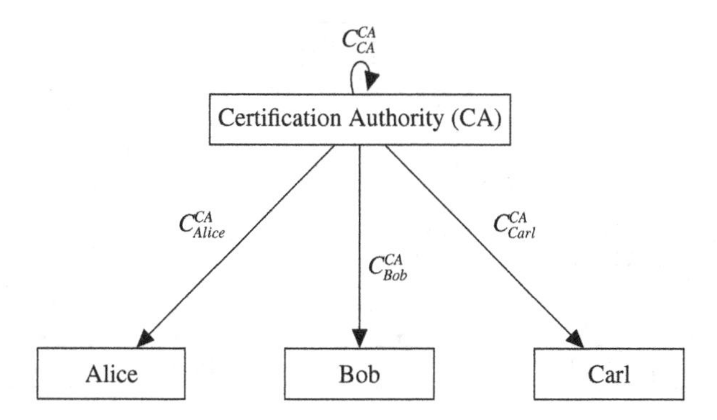

Fig. 3.9 A simple hierarchical PKI

certify. CAs make this process public in order for users to be able to determine the level of trust in the CAs. More details about this can be found in Chap. 8. Also, CAs assume liability for the authenticity of the public keys they certify. The most important example of a hierarchical PKI architecture is based on the X.509 standard. Issuers of X.509 certificates are assumed to be CAs.

Figure 3.9 shows a very simple hierarchical PKI. The participating entities are represented by rectangles while arrows represent certificates. In this PKI, a single CA has issued certificates to the end entities Alice, Bob, and Carl. The CA is the trust anchor. It has issued a so-called self-signed certificate to itself. This certificate contains the public key that is to be used to verify the signatures of the certificates issued by the CA. Both subject and issuer of the self-signed certificate are the CA itself. This certificate is depicted by a loop arrow from the CA to itself. All entities in the PKI establish direct trust in the trust anchor. Since the PKI users trust the trust anchor to sign certificates, the PKI users trust the authenticity of the public keys of Alice, Bob, and Carl. Also, if entities outside the PKI trust the trust anchor and its public key, then they also accept the public keys of Alice, Bob, and Carl as authentic. Note that the process of establishing trust in the trust anchor is very security-sensitive.

The PKI in Fig. 3.9 works as follows. Assume Carl wants to get an authentic copy of Alice's public key. Carl obtains the certificate issued by the trust anchor to Alice. Carl trusts the authenticity of the public key of the trust anchor directly. He verifies the CA's signature on Alice's certificate. Since Carl accepts the CA as a trusted authority that can authenticate public keys, the valid CA signature on Alice's certificate proves for Carl the authenticity of Alice's public key.

In general, a hierarchical PKI is a tree as shown in Fig. 3.10. The inner nodes are CAs. The leaves are the end entities. The root of the tree is called *root CA*. It is the trust anchor of the PKI. Each arrow from a node A to a node B represents a certificate issued by CA A to entity B. Frequently, there is a self-signed certificate of the trust anchor to itself.

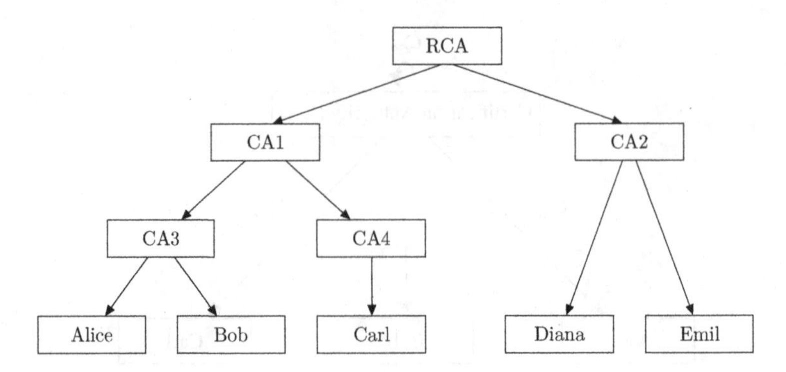

Fig. 3.10 A CA hierarchy

In a hierarchical PKI, trust in the authenticity of a public key is established via a *certification path*. Such a path is a finite sequence of certificates with the property that in all certificates except for the last one, the subject is the issuer of the subsequent certificate. For a more detailed discussion on verifying certification paths see Sect. 9.4. The path establishes trust in the public key of an entity if:

- The subject in the last certificate of the path is this entity and
- The issuer in the first certificate is the trust anchor.

As an example, we consider the PKI from Fig. 3.10. In this PKI, Alice trusts the public key of Diana. To establish this trust, she uses the certificate chain C^{RCA}_{RCA}, C^{RCA}_{CA2}, C^{CA2}_{Diana}.

The chain starts with the self-signed certificate of the root CA. In the next certificate, the root CA certifies the public key of CA2. In the third certificate, CA2 certifies the public key of Diana.

In principle, trust hierarchies of arbitrary depth are possible. However, trust may be weakened by introducing more depth in the hierarchy. Therefore, using trust hierarchies of large depth may not be a good idea. For example, the German signature law allows a trust hierarchy of depth 2 only. The trust anchor is a governmental agency which certifies commercial CAs, which in turn provide certificates to end entities.

3.3.1 Basic Constraints

The basic constraints extension of an X.509 certificate must always be set and be marked critical if the subject of the certificate is a CA. The ASN.1 description of this extension is shown in Listing 3.1. The extension has two fields. The first field, `cA`, is a Boolean value which is *true* if the certificate belongs to a CA and *false* otherwise. If this value is true, then the public key contained in the certificate can be used to verify signatures on certificates. Conversely, public keys that are used for verifying

```
BasicConstraints ::= SEQUENCE {
  cA                        BOOLEAN DEFAULT FALSE,
  pathLenConstraint         INTEGER (0..MAX) OPTIONAL }
```

Listing 3.1 Basic constraints extension

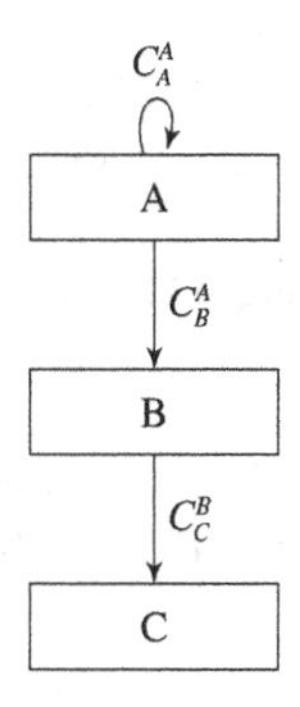

Fig. 3.11 PKI architecture—basic constraints

Table 3.1 Minimum values for pathLenConstraint

Certificate	Path length constraint
C_A^A	1
C_B^A	0
C_C^B	Not set

certificate signatures must be contained in certificates in which the cA field of the basic constraints extension is true.

The second field `pathLenConstraint` is used only for CA certificates in which the cA field is true and the `keyCertSign` bit is set in the key usage extension. The value of this field is an integer. It sets a limit on the number of intermediate CA certificates that may be found after this certificate in the certification path before the path is invalid. Self-issued certificates do not count. If such a limit is not desired then this field is empty.

For example, the PKI shown in Fig. 3.11 contains three certificates. The minimal pathLenConstraint values in the basic constraints extension of these three certificates still permitting a valid certification path are listed in Table 3.1.

3.4 Combining Trust Hierarchies

In practice, there exist many different independent hierarchical PKIs. Public administrations, companies, universities, and many other institutions operate such PKIs. The participants of each PKI are able to authenticate the public keys of the other users of this PKI. However, users of two different PKIs cannot authenticate their respective public keys unless the PKIs are combined. For example, in Fig. 3.12

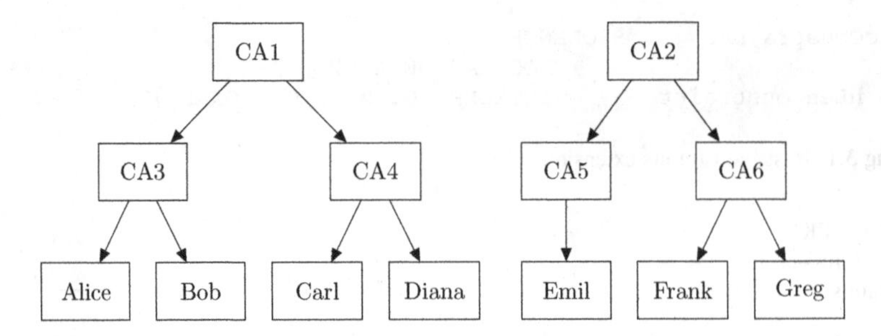

Fig. 3.12 Two independent PKIs

Alice cannot authenticate Emil's public key because Alice and Emil are participants of different PKIs. In this section we describe methods for combining PKIs.

3.4.1 Trusted Lists

A simple way for an entity to be able to authenticate the public keys of the end entities of several hierarchical PKIs is to accept the trust anchors of those PKIs as trustworthy. For this, membership in the PKI is not required. For example, in Fig. 3.12 Alice can accept CA2 as a new trust anchor. She is then able to construct certification paths starting at CA2. This implies that Alice trusts the public keys of Emil, Frank, and Greg although Alice is not a participant of their PKIs. An entity that wishes to use the public keys of the end entities of several hierarchical PKIs maintains a *trusted list* of root CAs. For example, this is done in Web browsers or certificate stores of operating systems. Those browsers and certificate stores come with a list of trusted CAs. Users can manage those lists by adding or removing CAs from the list.

A trusted list allows the owner of the list to use public keys of the end entities in the PKIs of which root CAs are contained in the list. But the trusted list does not enable the end entities of these PKIs to use the public key of the list owner unless he is a participant in all PKIs. This means that end entities of PKIs cannot send encrypted messages to the list owner and they cannot verify signatures of the list owner. Thus, trusted lists extend PKIs unilaterally only.

Figure 3.13 shows the certificate manager of the email client Thunderbird. While the certificates marked with *Builtin Object Token* are pre-installed, the certificates marked with *Software Security Device* were added by the user.

In addition to application-specific and end user-maintained trusted lists as seen in the example above, it is also possible to have trusted lists in an application-independent format. These lists may be maintained by a trusted party. A common way to achieve this is using the so-called *trust-service status lists* (TSLs) as defined in [6]. A TSL is a structured document (e.g. XML or PDF) containing

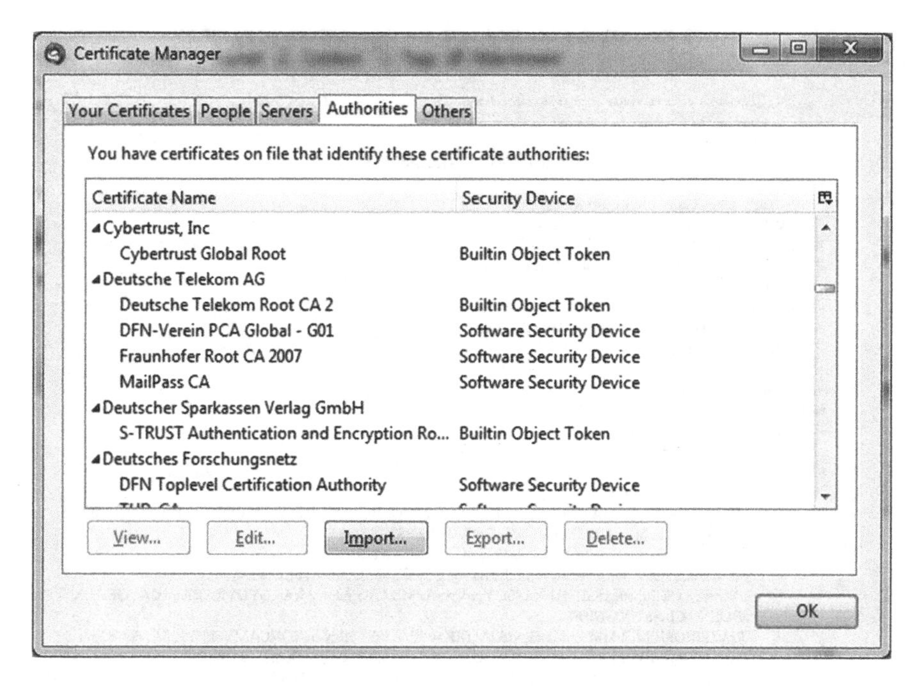

Fig. 3.13 Trusted list in Thunderbird

security-sensitive information on third parties which is signed by a trusted party. In the PKI context this information can be a list of trusted certificates. Figure 3.14 shows (a part of) a TSL issued by the European Commission containing, apart from other information, certain self-signed certificates of the member states.

3.4.2 *Common Root*

Another way to combine two or more hierarchical PKIs is to introduce a new *common root* CA as shown in Fig. 3.15. This means that each end entity of the combined PKIs replaces its original trust anchor by the new common root. As a consequence, certification paths that establish the authenticity of a public key have to be changed by prepending the common root. For example, in Fig. 3.15 the new certification path authenticating the public key of Emil for Alice is $C_{RCA}^{RCA}, C_{CA2}^{RCA}, C_{CA5}^{CA2}, C_{Emil}^{CA5}$.

If a common root is used, all users of all combined CAs can mutually authenticate their public keys. But all PKI users have to add the new common root or even replace their trust anchors by the new common root. This may cause problems. An institution may not be willing to accept an outside trust anchor or may find it to costly or error-prone to have all users modify their trust anchors.

Fig. 3.14 TSL of the European commission

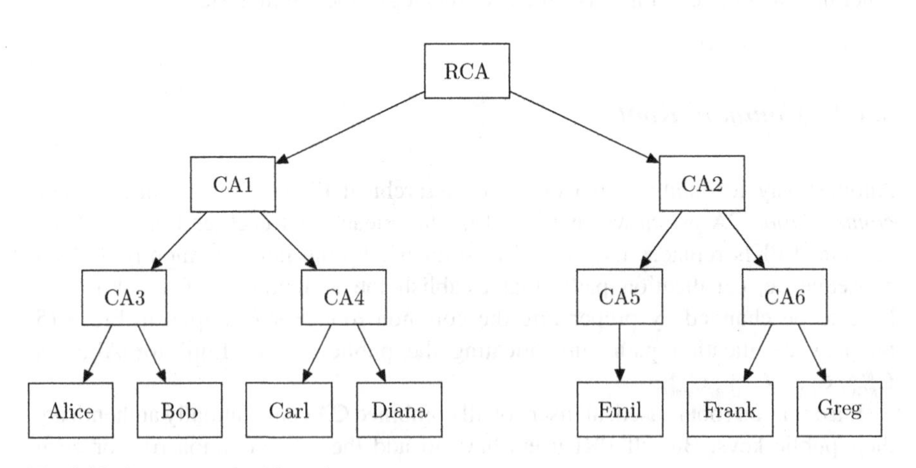

Fig. 3.15 A common root hierarchy

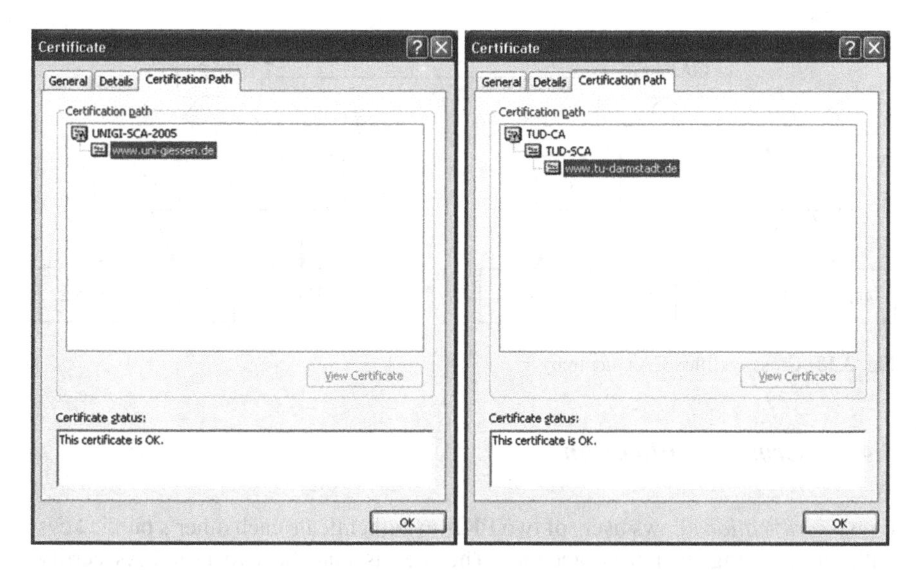

Fig. 3.16 Self-signed root certificates of two university PKIs

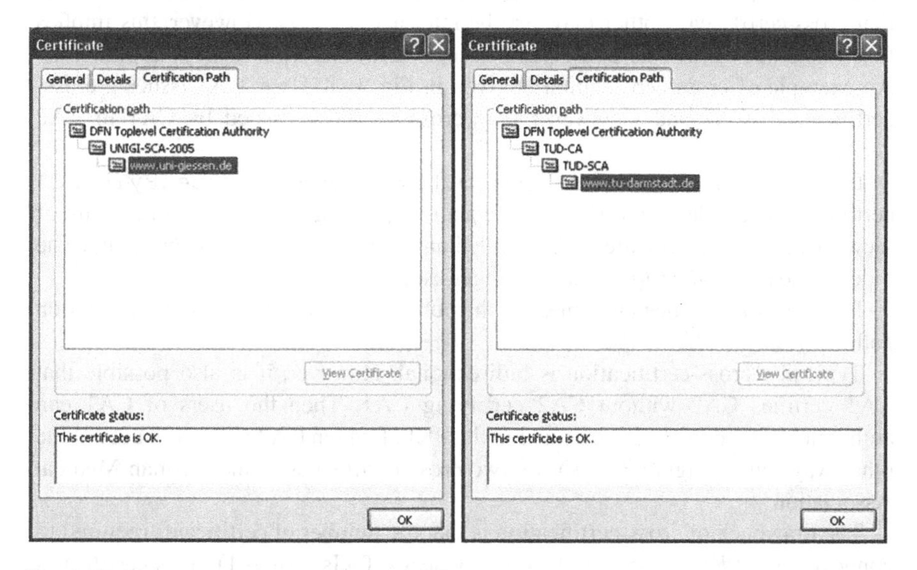

Fig. 3.17 Common root for two university PKIs

As an example, Fig. 3.16 shows the trust anchor certificates of the PKIs of the Universities of Gießen and Darmstadt in Germany. Figure 3.17 shows how the DFN root CA serves as a common root of these two PKIs.

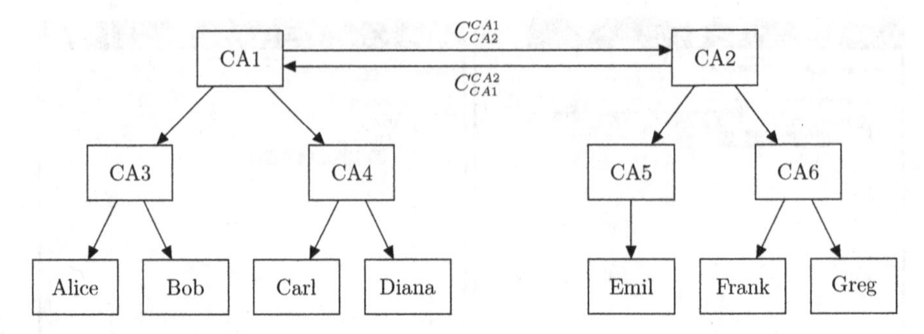

Fig. 3.18 Cross-certified CAs hierarchy

3.4.3 Cross-Certification

Cross-certification allows users of two PKIs to authenticate each other's public keys without replacing their trust anchors. The idea is that the two root CAs certify each other's public keys using so-called *cross-certificates*. In fact, the two CAs that cross-certify each other may also be intermediate CAs. However, this implies that only some of the PKI participants accept each other's public keys as authentic. An example of cross-certification is given in Fig. 3.18. Here, CA1 issues a cross-certificate to CA2 and vice versa. The cross-certificate issued by CA1 to CA2 is C^{CA1}_{CA2}.

From the point of view of CA1 this certificate is called an `issuedByThisCA` certificate. The older name *reverse certificate* is no longer used. From the point of view of CA2, the certificate C^{CA1}_{CA2} is called an `issuedToThisCA` certificate. The older name *forward certificate* is no longer used.

In the example, Alice authenticates the public key of Emil using the certification path $C^{CA1}_{CA1}, C^{CA1}_{CA2}, C^{CA2}_{CA5}, C^{CA5}_{Emil}$.

Typically, cross-certification is bidirectional. However, it is also possible that CA1 certifies CA2 without CA2 certifying CA1. Then the users of CA1 can authenticate the public keys certified in the PKI of which CA2 is the root but not the other way round. Figure 3.19 shows two cross-certificates of the German Medical Association.

The drawback of cross-certification is that the number of certificates required to connect many PKIs is large. In fact, to connect n PKIs, $n(n-1)$ cross-certificates must be issued. This may become impractical.

3.4.4 Bridge

Introducing *bridge CAs* allows users of several PKIs to authenticate each other's public keys without replacing their trust anchors. Also, compared to cross-certification much fewer new certificates are required when a bridge CA is introduced.

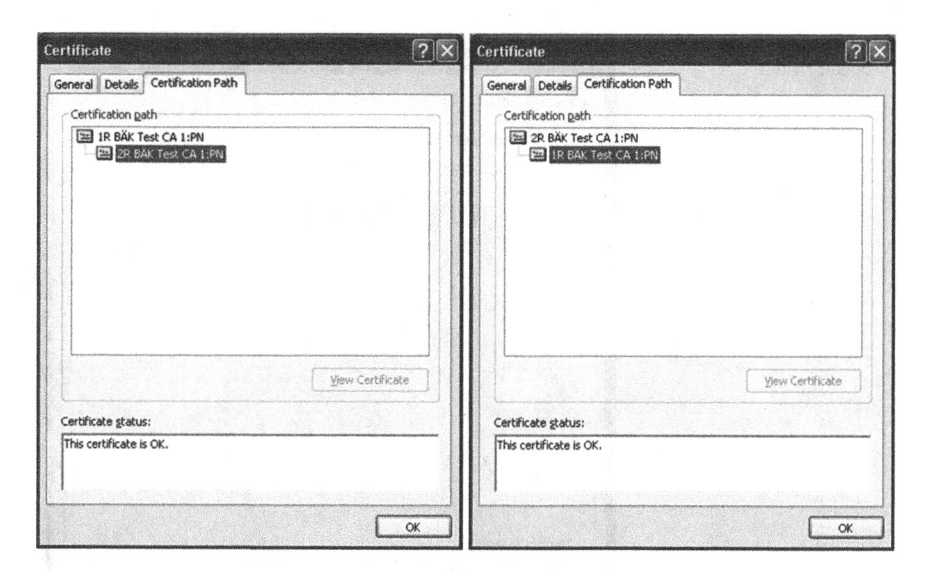

Fig. 3.19 Cross-certificates of the German medical association

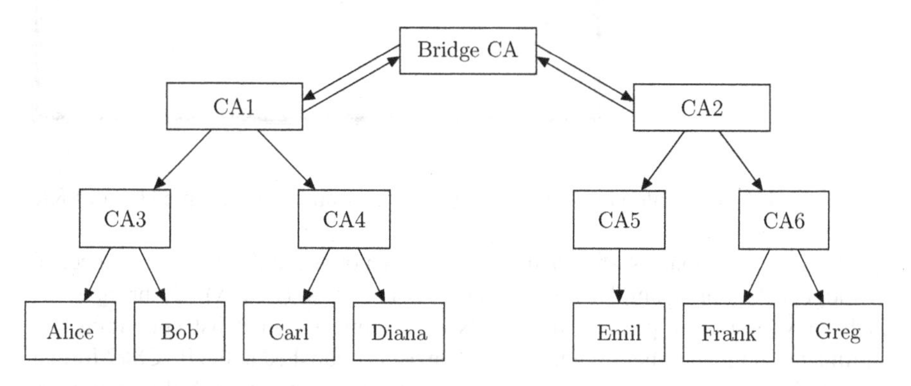

Fig. 3.20 A bridge CA hierarchy

The bridge is a new CA; however, it is not the new root for the PKIs to be combined. This is depicted in Fig. 3.20. The root CAs of the participating PKIs cross-certify the bridge CA and remain the trust anchors of their respective PKIs. Certification paths for authenticating public keys of end entities in the PKI of which the same CA is the root remain the same. Now consider Fig. 3.20 and assume that user Alice with trust anchor CA1 wishes to authenticate the public key of user Emil with trust anchor CA2. The corresponding certification path starts at CA1, the trust anchor for Alice, proceeds to the bridge, continues to CA2, and then follows the authentication path from CA2 to Emil. The path is C_{CA1}^{CA1}, $C_{Bridge\,CA}^{CA1}$, $C_{CA2}^{Bridge\,CA}$, C_{CA5}^{CA2}, C_{Emil}^{CA5}. It can be seen from this certification path that Alice does not need to update her trust anchor.

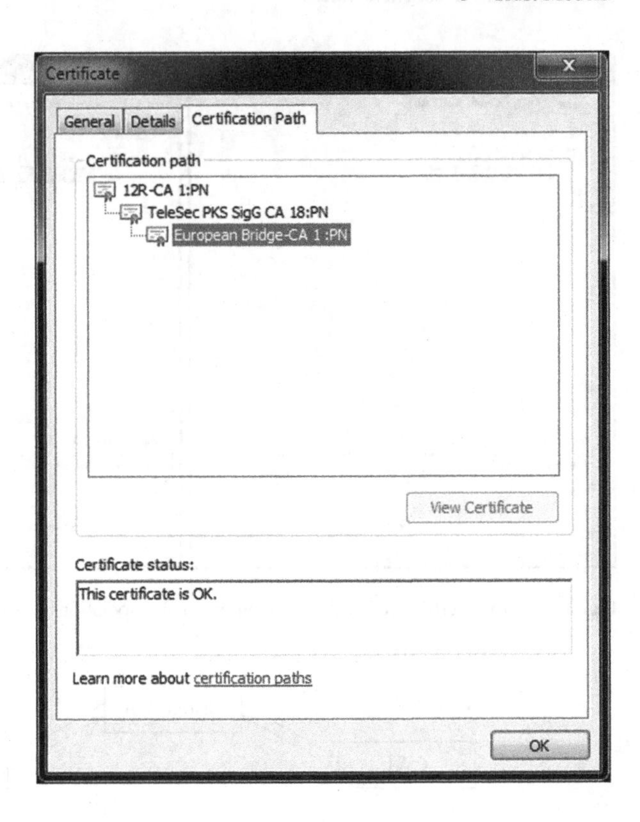

Fig. 3.21 Certificate of the European Bridge CA

The bridge CA is operated by a trustworthy organization that is accepted by all PKIs joining the bridge.

Cross-certification, as shown in Fig. 3.20, is not the only way to connect to a bridge CA. For example, the European Bridge CA (EBCA) [3] bridges the PKIs of various European organizations by issuing a signed list containing the certificates of the member organizations. The list is published in the PKCS#7 format (confer Sect. 7.5.1). Figure 3.21 shows the certificate of the EBCA that is used to sign the list. The certificate is subordinated to the German national root CA for legally binding electronic signatures, which makes the signature of the list legally binding.

3.5 Exercises

3.1. Enter the missing values into Table 3.2, which shows Alice's key ring.

3.2. Answer the following questions:

1. Where is the hash value of a complete X.509 certificate used?
2. Where is the hash value of the TBS part of an X.509 certificate used?

Table 3.2 Public key ring of Alice with missing values

Public key ring of Alice			
1		Public key owner	: Alice
		Owner trust/key legitimacy	: ultimate/complete
	1	Signer/trust in signer	: Alice/ultimate
2		Public key owner	: Bob
		Owner trust/key legitimacy	: marginal/
	1	Signer/trust in signer	: Alice/
	2	Signer/trust in signer	: Bob/
	3	Signer/trust in signer	: Carl/
3		Public key owner	: Carl
		Owner trust/key legitimacy	: none/
	1	Signer/trust in signer	: Bob/
	2	Signer/trust in signer	: Carl/
4		Public key owner	: Diana
		Owner trust/key legitimacy	: unknown/
	1	Signer/trust in signer	: Bob/
	2	Signer/trust in signer	: Carl/
	3	Signer/trust in signer	: Diana/
	4	Signer/trust in signer	: Emil/
	5	Signer/trust in signer	: Frank/
5		Public key owner	: Emil
		Owner trust/key legitimacy	: marginal/
	1	Signer/trust in signer	: Alice/
	2	Signer/trust in signer	: Bob/
	3	Signer/trust in signer	: Emil/
6		Public key owner	: Frank
		Owner trust/key legitimacy	: marginal/
	1	Signer/trust in signer	: Alice/
	2	Signer/trust in signer	: Frank/

3.3. Consider the hierarchical PKI in Fig. 3.22.

1. What are the possible trust anchors for Alice that enable her to trust the authenticity of Bob's public key?
2. What are the possible trust anchors for Alice that enable her to trust the authenticity of Carl's public key?
3. Is it possible for Alice to be convinced of the authenticity of Emil's public key without trusting the Dep2 CA?
4. Is it possible for Alice to be convinced of the authenticity of Emil's public key without trusting the Org2 CA?

3.4. Consider the PKIs in Fig. 3.23.

1. How can these five PKIs be connected in such a way that any end entity can trust the public key of any other end entity?
2. How many new certificates are issued in each of the solutions?
3. How many trust anchors do the end entities have in each case? What are the trust anchors of end entity G in each case?

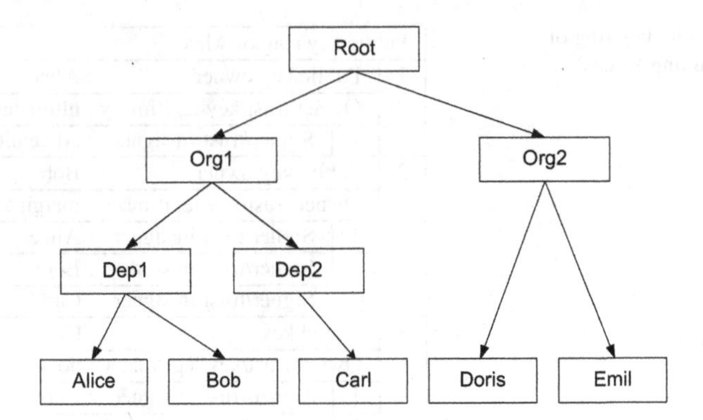

Fig. 3.22 A hierarchical PKI

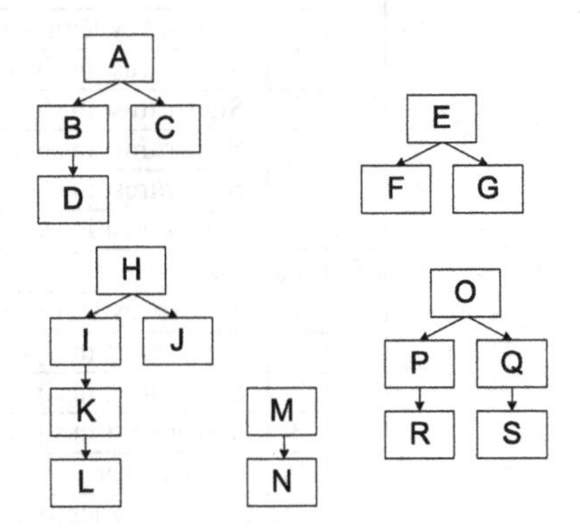

Fig. 3.23 A PKI setup

References

1. CAcert – root keys, http://www.cacert.org/index.php?id=3
2. J. Callas, L. Donnerhacke, H. Finney, D. Shaw, R. Thayer, OpenPGP message format, in *IETF Request for Comments*, 4880, Nov 2007
3. EBCA – TeleTrusT European Bridge CA, https://www.ebca.de/en/
4. GNU Privacy Guard, http://www.gnupg.org/
5. J. Jonczy, M. Wüthrich, R. Haenni, A probabilistic trust model for GnuPG, in *Proceedings of the 23rd Chaos Communication Congress, 23C3*, Berlin, pp. 61–66, 2006
6. The European Telecommunications Standards Institute (ETSI), Electronic signatures and infrastructures (ESI); Provision of harmonized trust-service status information. ETSI TS 102 231 V3.1.2, Dec 2009
7. The International PGP Home Page, http://www.pgpi.org/
8. T. Ylonen, C. Lonvick, The secure shell (SSH) protocol architecture, in *IETF Request for Comments*, 4251, Jan 2006

Chapter 4
Private Keys

In the previous chapters we have seen how certificates and trust models can be used to authenticate public keys. A further task of public key infrastructures is to support the users in keeping their private keys secret. In this chapter we show how this can be achieved. We introduce a private key life cycle model and discuss software- and hardware-based solutions for storing and protecting private keys.

4.1 Private Key Life Cycle

We begin our discussion by examining the life cycle of a private key, shown in Fig. 4.1, as presented in [10]. There, the life cycle is described using the terminology of finite state machines. This means that a private key can be in different states and there are transitions between the states. The possible states are depicted by circles, the transitions are symbolized by arrows pointing from a source state to a destination state.

We summarize the possible states and transitions. Initially, the key is *non-existent*. From this state the only possible transition is to *generate* the key after which it is *storable*. This means that the key exists in some volatile memory waiting to be processed further. Key generation has to be implemented securely. For example, random number generation must be done in such a way that the random numbers cannot be predicted. Storable keys permit two transitions: they can be *stored* in some personal security environment (PSE) after which they are *deliverable* to the potential users. The PSE is protected against unauthorized access, for example by a pass-phrase. In some cases it might be necessary to *restore* a deliverable key from the PSE to the storable state. When keys are storable it is also possible to *deposit* them, after which they are in the *archived* state. For example, this may be necessary to back up decryption keys. Archived keys can be *recovered*, after which they are storable again. Deliverable keys are *delivered* to their owners after which they are *usable* for signing or decryption. If the keys do not reach the user, they may have to be *retracted*. In all states in which keys exist, they may be *copied* or *destroyed*.

J.A. Buchmann et al., *Introduction to Public Key Infrastructures*,
DOI 10.1007/978-3-642-40657-7_4, © Springer-Verlag Berlin Heidelberg 2013

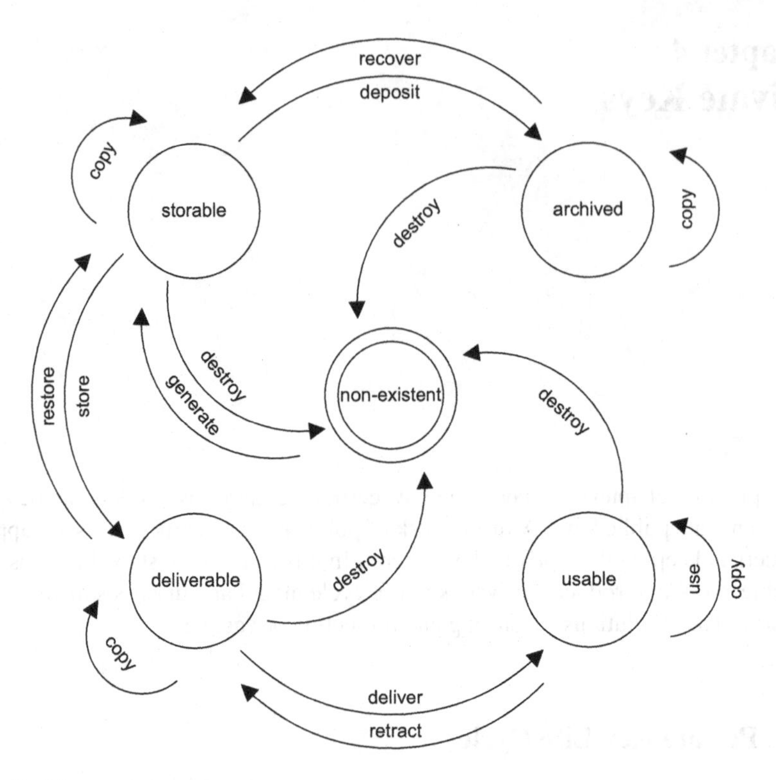

Fig. 4.1 The private key life cycle

The latter may happen to keys that are expired and do not need to be archived. Destruction of keys is a sensitive issue which is discussed in [4] and [1].

In their different states the private keys may have additional attributes. These attributes and the policies that are applied in the PKI determine which transitions are selected. For example, a compromised signature key may be destroyed immediately; a compromised decryption key may be required to decrypt old data and may therefore be destroyed later.

4.2 Personal Security Environments

Personal security environments (PSEs) protect private keys of users against unauthorized access. Before a PSE grants access to its data, it requires authorization from the accessing party. Typically, authorization is based on a personal identification number (PIN) or a pass-phrase, which are only known to authorized entities.

Figure 4.2 presents an overview of widespread PSE types. PSEs can be implemented in software (for example, encrypted files) or in hardware (for example, smart cards). While software PSEs provide better flexibility and platform independence,

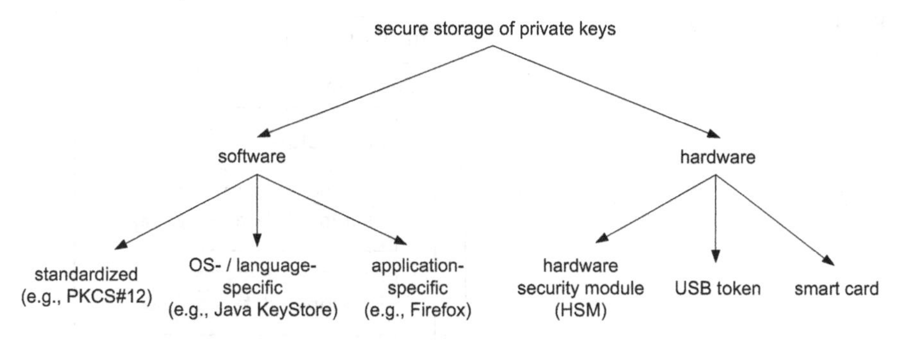

Fig. 4.2 Overview of personal security environments

hardware PSEs have a higher security level. The following sections describe commonly used PSEs in more detail.

4.3 Software PSEs

4.3.1 PKCS#12

The PKCS#12 [8] standard specifies a format that allows storing and transferring security-sensitive data such as private keys and certificates. Many applications, for example Web browsers and operating systems, can process private keys stored in this format. Typically, they import the PKCS#12 files, extract the keys, and store them using some internal representation.

Figure 4.3 shows the structure of PKCS#12. It is an `AuthenticatedSafe` that contains one or more `ContentInfo` structures. It protects the authenticity and confidentiality of its contents using a secret password or a public key technique.

We describe the PKCS#12 structure in more detail.

AuthenticatedSafe The content that needs to be protected is placed in a container called AuthenticatedSafe. It supports the following two authentication modes.

Public Key Integrity Mode In this mode the AuthenticatedSafe is digitally signed using a private key of the content originator. The corresponding public key is appended to the data. Receivers may check the authenticity of the content by verifying the signature and by authenticating the public key, for example, by obtaining its fingerprint from the content owner.

Password Integrity Mode In this mode a MAC value of the AuthenticatedSafe's content is calculated using a symmetric key which is derived from a password, some salt bits, and a counter. The salt and the counter are appended to the data. Recipients who know the password can derive the same key and authenticate the content. The salt bits and the counter prevent replay attacks.

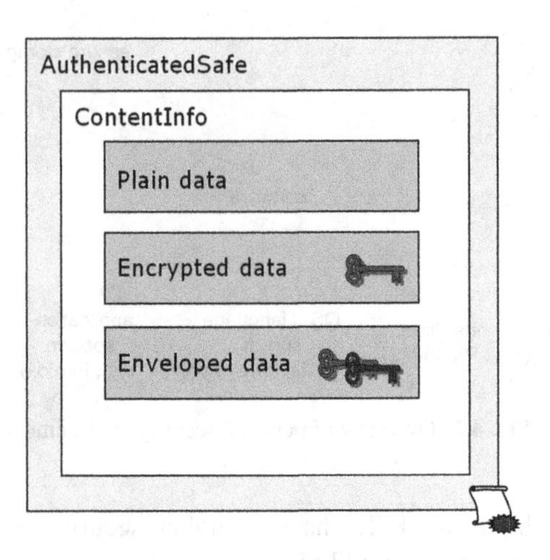

Fig. 4.3 The PKCS#12 format

ContentInfo An AuthenticatedSafe may contain several ContentInfo structures. Three ContentInfo types are possible: *data*, *enveloped*, and *encrypted*. The (plain) data type is used if no confidentiality protection is required. The other two types are used in the modes explained now.

Public Key Privacy Mode In the public key privacy mode the type enveloped data is used. The content is encrypted using a symmetric encryption key which in turn is encrypted with the public key of the receiver. The receiver is able to decrypt the encrypted symmetric key and the protected content.

Password Privacy Mode In the password privacy mode the type encrypted data is used. The content is encrypted with a symmetric key which is derived from a password, additional salt bits and a counter. The salt and the counter are appended to the data. The password must be known to the recipient of the file to decrypt the data.

Combinations If both a confidentiality and an authenticity protecting mechanism are used, four combinations of the modes are possible. Most commonly, the password privacy and password integrity modes are used and the same password is applied in both modes. Most operating systems and applications are able to process this format and extract private keys from it.

Typically, for PKCS#12, files, the extensions `.p12` or `.pfx` are used. This allows their automatic processing, for example, as email attachments. A critical discussion of the PKCS#12 standard can be found in [3].

4.3.2 PKCS#8

Another format to store private keys is specified in the PKCS#8 standard [7]. This format allows storing private keys in encrypted and non-encrypted form. However, in practice this format is almost never used by itself for key transport. Instead, a PKCS#8-encoded private key is transported by placing it into the file content of a PKCS#12 file. If the PKCS#12 privacy mode is used, the PKCS#8 file needs no encryption.

4.3.3 *Java KeyStore*

Java KeyStore is a Java class that supports storing, accessing, and modifying data such as private keys, secret keys, and certificates. Also, this class supports the interaction with standard formats such as PKCS#12 (see Sect. 4.3.1) and PKCS#11 (see Sect. 4.4.3). Using the KeyStore class it is possible to implement application-specific Java key stores that are called *key store types*. Listing 4.1 shows how to choose and use KeyStores in a Java program. Java has the following predefined key store types.

JKS This is a very simple built-in key store implementation. It is the default type. It implements the key store as a single file using a proprietary format. According to the Java documentation it applies some kind of weak encryption and integrity protection to the stored keys.

JCEKS This type is provided as an alternative to the JKS type. It also implements the KeyStore using a proprietary single file format. According to the Java documentation this type provides stronger cryptography for encryption and integrity protection. The default encryption method is password-based Triple-DES.

PKCS12 This type allows creating, accessing, and modifying standard PKCS#12 files using the KeyStore interface.

PKCS11 This type allows accessing and modifying PKCS#11-based hardware PSEs using the KeyStore interface. It relies on PKCS#11 libraries that are usually delivered with the PSE hardware.

4.3.4 *Application-Specific Formats*

There are also many application-specific formats for storing private keys.

Microsoft Windows Figure 4.4 shows the certificate and key manager of the Microsoft Windows operating systems. It can be opened by executing `certmgr.msc` on the command line. It stores keys and certificates in several categories. When a private key is imported it is stored together with the corresponding certificate in the selected category. The certificate manager provides an API to allow applications to access the store and a GUI for the user. Figure 4.4 shows the GUI with the categories on the left side and the certificates stored in the "Third-Party Root Certification Authorities" category.

```
import java.io.FileInputStream;
import java.io.IOException;
import java.security.Key;
import java.security.KeyStore;
import java.security.KeyStoreException;
import java.security.NoSuchAlgorithmException;
import java.security.PrivateKey;
import java.security.UnrecoverableKeyException;
import java.security.cert.CertificateException;

public class KeyStoreUse {

  public static void main(String[] args) throws KeyStoreException
      , NoSuchAlgorithmException, CertificateException,
      IOException, UnrecoverableKeyException  {

    KeyStore ks = KeyStore.getInstance("JKS");

    /* replace the previous line with this one to work with JCEKS
        KeyStores */
    /* KeyStore ks = KeyStore.getInstance("JCEKS"); */

    /* replace the previous line with this one to work with
        PKCS12 (since Java version 5.0) */
    /* KeyStore ks = KeyStore.getInstance("pkcs12"); */

    char[] password = "the_password_that_protects_the_file".
        toCharArray();
    FileInputStream fis = new FileInputStream("path/to/file.ks");
    ks.load(fis, password);
    fis.close();

    Key key = ks.getKey("alias", password);
    PrivateKey privKey = (PrivateKey) key;

    // perform an operation like signing with this private key

  }
}
```

Listing 4.1 How to use a Java KeyStore

Thunderbird The email client Thunderbird has its own certificate and key manager which stores keys and certificates in several categories. Imported private keys are stored together with the corresponding certificate in the chosen category. Figure 4.5 shows Thunderbird's certificate manager. It shows the different categories as tabs at the top of the window. The category "Your Certificates" is opened and the certificates belonging to this category are visible. The certificate manager also provides an API allowing plug-ins and add-ons to access the store.

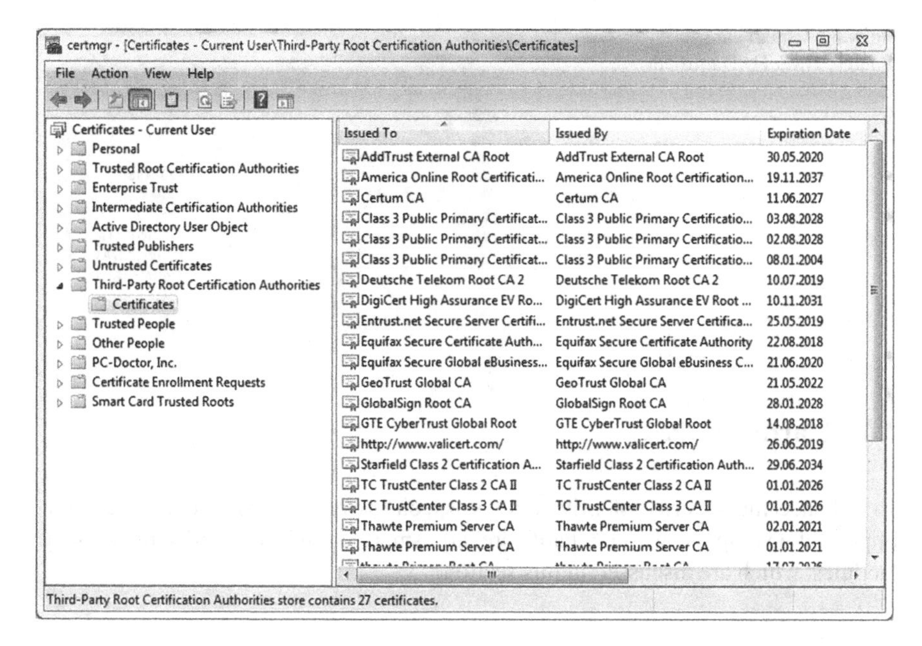

Fig. 4.4 Microsoft Windows certificate manager

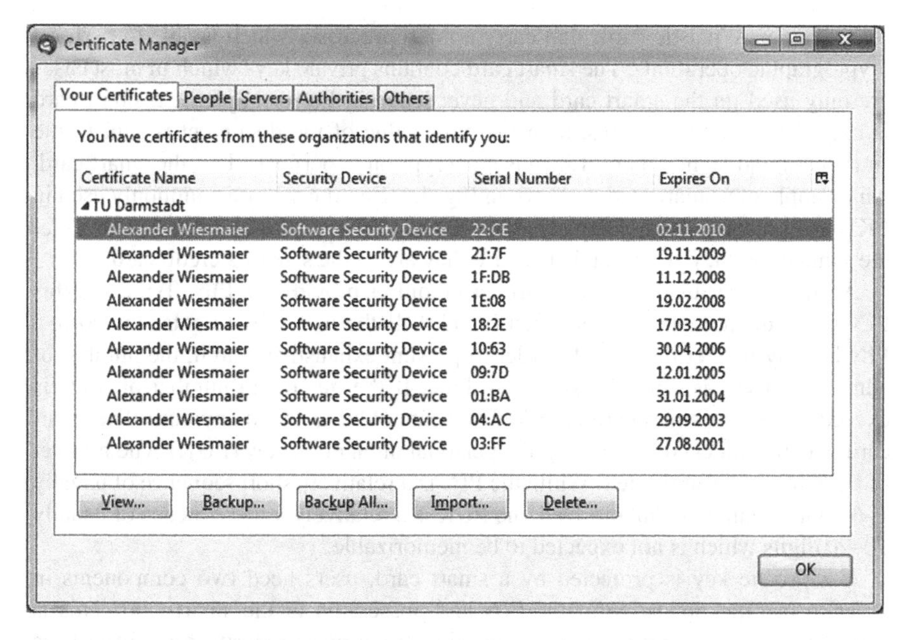

Fig. 4.5 Thunderbird's certificate manager

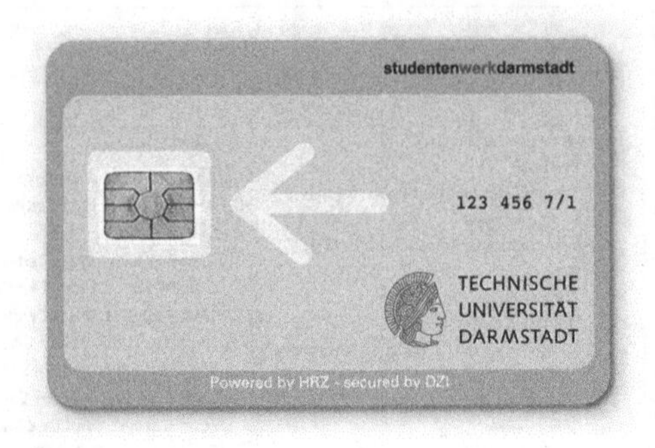

Fig. 4.6 A smart card (Source: HRZ TU Darmstadt)

4.4 Hardware PSEs

To enhance the security of private keys, special-purpose hardware can be used. Important examples of such hardware are smart cards and hardware security modules, which are discussed in this section.

4.4.1 Smart Cards

Smart cards are plastic cards that carry a microprocessor which is able to perform cryptographic operations. The smart card contains private keys which in most cases are only used on the smart card and never leave it. For example, if a smart card is used for digital signatures, then the hash value of the document is sent to the smart card and signed there. Likewise, decryption is performed on the smart card. An example of a smart card is shown in Fig. 4.6. Smart cards are commonly used in PKIs as hardware PSEs. They provide very secure storage of private keys, and, at the same time, they are portable because they are of the size of a credit card.

Access to private keys on a smart card requires providing a PIN. Typically, this PIN is a sequence of digits between 0 and 9. If the user fails to enter the correct PIN he may reenter the PIN. In order to prevent exhaustive search, the number of admissible trials is limited, usually to three. If the maximum number of trials is exceeded then the smart card is locked and cannot be accessed anymore. The smart card may be unlocked by entering the personal unblocking key (PUK). The number of PUK trials is also limited. While the PIN is a relatively short sequence of usually 4–6 digits meant to be memorized, the PUK is a relatively long sequence of usually 10–20 digits which is not expected to be memorizable.

If a private key is protected by a smart card, users need two components in order to use it: knowledge of the PIN and possession of the smart card. This is one of the reasons why smart cards provide much more security for private keys than software PSEs. Since using the private key requires the possession of the smart card, it can only be used by exactly one person at a time. Also, since the use of the smart card requires the knowledge of the correct PIN, only the owner of the smart card can use it. On some smart cards the PIN authorization is replaced by a biometric authorization. That enhances the security level of private keys even further, as biometric attributes cannot be passed on to other persons.

Most smart cards run a proprietary operating system that manages its resources. For example, it prevents private keys from being read from the smart card. It is stored in the read-only memory (ROM) of the card and cannot be changed. Keys, certificates, and other data are written in the electrically erasable programmable read-only memory (EEPROM). In order to change or erase this information, special access control mechanisms are used.

There are also Java-enabled smart cards, so-called *Java cards*. In addition to their proprietary operating systems, these smart cards have a *Java card runtime environment* (JCRE) and can process *Java card applets*. Such applets are usually written in a reduced version of the Java programming language. Java cards are easier to program than other smart cards since they use a standard high-level programming language. However, if a cryptographic algorithm is not available on the smart card processor, then its execution in a Java card is slow.

Frequently, smart cards are evaluated according to some evaluation framework such as information technology security evaluation criteria (ITSEC) or common criteria. This evaluation provides some evidence of the security of the smart card.

4.4.2 Smart Card Readers

In order to operate a smart card, a smart card reader is required which is connected to a host computer. The reader handles the communication between the host and the smart card. There are contact smart cards, contactless smart cards, and dual interface cards. Contact smart cards use a metal contact to transmit data between the reader and the smart card. Contactless smart cards use radio frequencies to transmit data. Dual interface cards have both a contact and a contactless interface.

In addition to simple card readers that can only read from the card and write on the card, there are readers with key pads for entering the PIN. The advantage of such readers is that the PIN remains unknown to the host computer. This prevents malicious software from eavesdropping and abusing the PIN. There are also smart card readers with PIN pad and display. The display can be used to enhance the security even further. It can be used to display messages to the user. For example, the display may show a hash or even the whole document that is to be signed.

There are also programmable readers called *FINREAD*. These readers are programmed by applets written in the Java language. This applet is called a finlet. These readers can be customized for several applications. For example, they may show the amount of an electronic banking transaction before the transaction is signed.

Finally, the smart card may be placed in a USB token. Then the smart card communicates with its host via the USB interface. This solution has several advantages. It may be cheaper since no extra card reader is required. More importantly, such

```
CK_SLOT_ID slotID;
CK_ULONG ulCount;
CK_MECHANISM_TYPE_PTR pMechanismList;
CK_RV rv;
...
rv = C_GetMechanismList(slotID, NULL_PTR, &ulCount);
if ((rv == CKR_OK) && (ulCount > 0)) {
    pMechanismList = (CK_MECHANISM_TYPE_PTR) malloc(ulCount*
         sizeof(CK_MECHANISM_TYPE));
    rv = C_GetMechanismList(slotID, pMechanismList, &ulCount);
    if (rv == CKR_OK) {
        ...
    }
    free(pMechanismList);
}
```

Listing 4.2 PKCS#11 example

a token can be used with almost every host computer since modern computers are equipped with a USB interface.

4.4.3 Smart Card Communication Interfaces

In this section we explain three programming interfaces for the communication between hosts and smart cards: PKCS#11, PC/SC, and CT-API. These standards are supported by all major operating systems such as Microsoft Windows, Mac OS, and Linux.

PKCS#11 The PKCS#11 standard [9] specifies an interface to cryptographic tokens such as smart cards that are able to store, manage, and use cryptographic keys. PKCS#11 is the de facto standard for accessing smart cards. It defines a platform-independent high-level API called *cryptographic token interface* (cryptoki) for accessing cryptographic hardware.

The API is written in the programming language C and defines data types and functions for the most common cryptographic objects (such as keys and certificates) and mechanisms (such as encryption and hashing). Many of the available cryptographic hardware tokens exceed the functionality defined in PKCS#11. Usually, the manufacturers provide proprietary extensions to the PKCS#11 standard that enable programmers to use the full potential of their hardware. Listing 4.2 shows sample C code for requesting the list of supported mechanisms (algorithms) from a token. The example is taken from [9].

Smart card manufacturers usually provide PKCS#11 implementations for their cards. Correspondingly, clients have to provide support for PKCS#11 if they wish to support smart cards. For example, Firefox and Thunderbird can access keys and certificates on smart cards using PKCS#11. Figure 4.7 shows the dialog window for installing a PKCS#11 library, usually provided in the form of a `.dll` file for

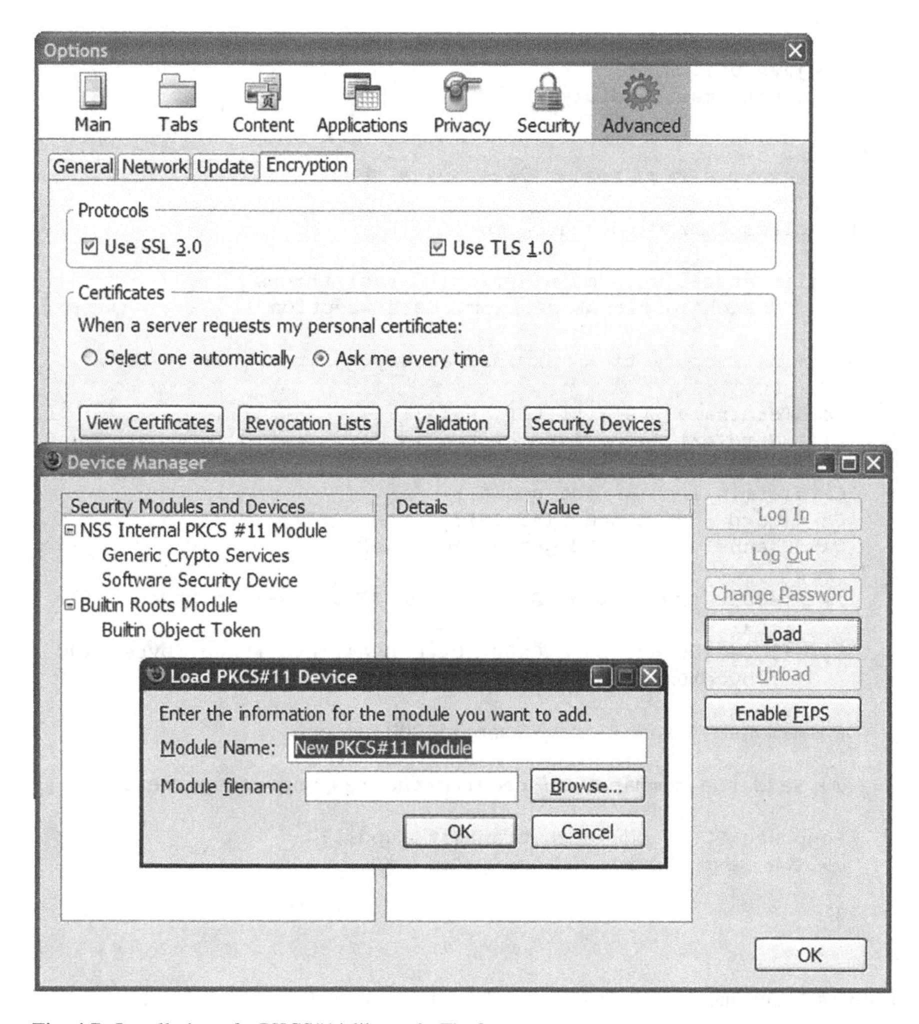

Fig. 4.7 Installation of a PKCS#11 library in Firefox

Windows or `.so` file for Linux, in Firefox. Other applications are also based on PKCS#11, for example authentication modules for UNIX or Microsoft Windows.

PC/SC The interoperability specification for ICCs and personal computer systems (PC/SC) interface is described in [6]. It is a widely accepted standard for smart cards and readers. Most available card readers are delivered with a PC/SC driver. PC/SC does not provide a high-level programming interface like PKCS#11, but requires the programmer to send binary commands to the card. In Listing 4.3 we show an example of a communication with a smart card via PC/SC.

CT-API CT-API (card terminal-API) is widely used in Germany. It is specified in [2]. This interface is very simple. It only specifies three different functions, one for

```
import java.security.NoSuchAlgorithmException;
import java.util.List;
import javax.smartcardio.*;

/**
 * This example requires at least Java version 6.
 */
public class SmartCardConnection {

    public static void main(String[] args) throws
         NoSuchAlgorithmException, CardException {

    TerminalFactory tf = TerminalFactory.getInstance("PC/SC",
         null);
    CardTerminals cts = tf.terminals();
    List<CardTerminal> list  = cts.list();

    CardTerminal ct = list.get(0);
    Card card = ct.connect("T=1");
    CardChannel cc = card.getBasicChannel();

    /* Prepare the APDU 00 A4 00 0C 02 3F 00 (select Master File)
         . */
    byte[] select = {(byte)0x00,(byte)0xA4,(byte)0x00,(byte)0x0C
         ,(byte)0x02,(byte)0x3F,(byte)0x00};

    CommandAPDU capdu = new CommandAPDU(select);

    /* Send the command and receive the response from the card.
         */
    ResponseAPDU rapdu = cc.transmit(capdu);
    /* The response should be 90 00. */

    }
}
```

Listing 4.3 Communication with a smart card over PC/SC using Java

opening the connection to the terminal or the card (`CT_init`), one for sending and receiving data (`CT_data`), and one for closing the previously opened connection (`CT_close`). Although CT-API is simple it allows us to perform all necessary functions with a card and the card reader by sending binary commands.

4.4.4 Hardware Security Module

A hardware security module (HSM) is a hardware device that is optimized for cryptographic operations without the size constraints of a smart card. HSMs are available as PCI devices, as external devices that can be connected to a PC, for example via USB, and as network devices which can be accessed using TCP/IP. HSMs are operated using the PKCS#11 standard or the cryptographic architectures JCA/JCE of the Java programming language. Typical HSM operations are secure (pseudo)random number generation, key pair generation, calculation of digital signatures, encryption, and decryption. Frequently, HSMs also support cryptographic protocols such as TLS. HSMs are commonly used in server systems which have to perform many cryptographic operations in a short period of time. Example applications are time-stamping servers, online certificate status protocol servers, and server-based certificate validation protocol servers. HSMs are typically protected against mechanical, temperature, electrical, electronic, and chemical attacks and support multi-person control.

Most HSMs are evaluated according to the FIPS 140-1 and FIPS 140-2 standards [5]. These standards define requirements for cryptographic modules and have different levels of assurance. Other evaluation frameworks such as the common criteria are also used to evaluate HSMs.

4.5 Exercises

4.1. Answer the following questions:

1. What modes of content protection are allowed by PKCS#12?
2. How can the modes be combined?
3. What secrets must be known and to whom in each mode?

4.2. Describe the steps that are executed when a signature of a document is generated on a smart card that is connected to a PC via a smart card reader.

References

1. G. Di Crescenzo, N. Ferguson, R. Impagliazzo, M. Jakobsson, How to forget a secret, in *Proceedings of the 16th Annual Symposium on Theoretical Aspects of Computer Science (STACS'99)*, Trier, ed. by C. Meinel, S. Tison. Number 1563 in Lecture Notes in Computer Science (LNCS). (Springer, Berlin/Heidelberg, 1999), pp. 500–509
2. DT, FHG, SIT, TÜV, TELETRUST, CT-API 1.1, Application independent CardTerminal application programming interface for ICC applications (2002), http://www.tuvit.de/cps/rde/xbcr/tuevit_de/CTAPI11EN.pdf
3. P. Gutmann, PFX – How not to design a crypto protocol/standard, http://www.cs.auckland.ac.nz/~pgut001/pubs/pfx.html
4. P. Gutmann, Secure deletion of data from magnetic and solid-state memory, in *Proceedings of the 6th USENIX Security Symposium*, San Jose, July 1996
5. NIST, FIPS standards, http://csrc.nist.gov/publications/PubsFIPS.html
6. PCSC Workgroup, PC/SC specification, http://www.pcscworkgroup.com/
7. RSA Laboratories, PKCS #8 v1.2: private-key information syntax standard (1993), http://www.rsa.com/rsalabs/node.asp?id=2130

8. RSA Laboratories, PKCS #12 v1.0: personal information exchange syntax (1999), http://www.rsa.com/rsalabs/node.asp?id=2138
9. RSA Laboratories, PKCS #11 v2.20: cryptographic token interface standard (2004), http://www.rsa.com/rsalabs/node.asp?id=2133
10. A. Wiesmaier, Secure private key management in adaptable public key infrastructures. Ph.D. thesis, Cryptography and Computer Algebra Group, Technische Universität Darmstadt. (Mensch und Buch Verlag, Berlin, 2009). ISBN-13: 978-3-86664-646-9

Chapter 5
Revocation

The validity period of certificates may be quite long. For example, X.509 SSL server certificates are typically valid for at least 2 years. However, it may happen that during the validity period a certificate has to be invalidated, for example, if the private key that corresponds to the public key in the certificate has been compromised. The process of invalidating the certificate before its expiration time is called *revocation*. In this chapter, we discuss revocation and strategies to publish revocation information.

5.1 Requirements

If a certificate is revoked, revocation information is generated and made available. The certificate revocation information should have several properties. It must contain the revoked certificates and the revocation time. In addition, it may also contain the reason for the revocation. This may be useful to estimate potential damage caused by using a revoked certificate. The revocation information must be anywhere and always available since it cannot be predicted where and when the public key contained in the certificate is used. The authenticity of the revocation information must be verifiable by anyone. The revocation information must be as fresh as possible. There may always be a delay between the revocation of a certificate and the publication of the revocation information. However, this delay should be as short as possible.

A more detailed analysis of revocation requirements as well as a list of criteria for evaluating revocation mechanisms can be found in [1].

J.A. Buchmann et al., *Introduction to Public Key Infrastructures*,
DOI 10.1007/978-3-642-40657-7_5,

```
CertificateList ::= SEQUENCE {
  tbsCertiList           TBSCertList,
  signatureAlgorithm     AlgorithmIdentifier,
  signatureValue         BIT STRING  }

TBSCertList  ::=  SEQUENCE  {
  version                Version OPTIONAL,
  signature              AlgorithmIdentifier,
  issuer                 Name,
  thisUpdate             Time,
  nextUpdate             Time OPTIONAL,
  revokedCertificates    SEQUENCE OF SEQUENCE  {
    userCertificate        CertificateSerialNumber,
    revocationDate         Time,
    crlEntryExtensions     Extensions OPTIONAL }  OPTIONAL,
  crlExtensions          [0]  EXPLICIT Extensions OPTIONAL }
```

Listing 5.1 ASN.1 specification of an X.509 CRL

5.2 Certificate Revocation Lists

One way to publish revocation information is to use *certificate revocation lists* (CRLs). A CRL is a list of revoked certificates which is digitally signed to prove its authenticity. CRLs are regularly updated. When a CRL is updated, newly revoked certificates are inserted into the CRL. There are *direct CRLs* and *indirect CRLs*. Direct CRLs only contain certificates of one issuer and are issued and signed by that issuer. In contrast, an indirect CRL may contain certificates of several issuers and is signed by the so-called CRL issuer.

Users who wish to obtain revocation information download the CRL and verify its digital signature. Then they check whether the certificate that they are interested in is contained in the CRL.

One problem with CRLs is that they may become quite large since expired certificates are not always removed from the CRL. Therefore, *delta CRLs* have been introduced which only contain the certificates that have been revoked after the publication of the last *full CRL*. The full CRL contains all revoked certificates. It is also referred to as *complete CRL*.

CRLs are specified in the X.509 standard. The ASN.1 description of an X.509 CRL is shown in Listing 5.1. Its structure is presented in Fig. 5.1. This structure resembles that of an X.509 certificate.

In the remainder of this section we explain X.509 CRLs in more detail.

5.2.1 Basic Fields

We describe the basic fields of a CRL.

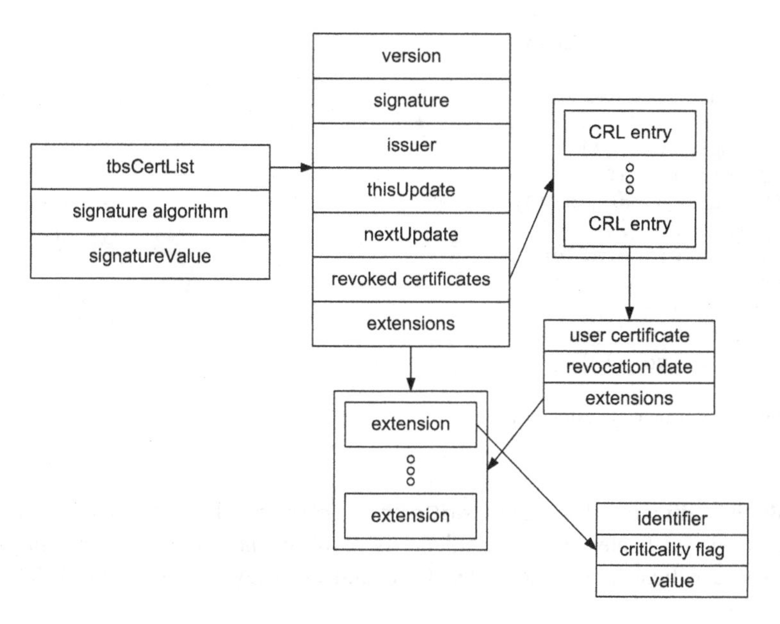

Fig. 5.1 The structure of an X.509v2 CRL

version The *version* field contains the version of the CRL specification that is used for this CRL. Currently, only version 2 CRLs are used. They support CRL extensions. Therefore, the entry of this field is always 2.

signature The *signature* field contains the OID of the signature algorithm that has been used to sign the CRL. This OID must be the same as the OID that is contained in the part of the CRL that is not signed.

issuer The *issuer* field contains the distinguished name of the CRL issuer who has signed the CRL. This field must not be empty.

thisUpdate The content of the `thisUpdate` field is the time and date when the CRL was issued. It informs users that certificates not included in the CRL were not revoked before this point in time. The individual revocation time of a certificate is contained in the corresponding *CRL entry*.

nextUpdate The `nextUpdate` field contains the date and time when the next update of the CRL will be available. This field is optional. If it is present, the follow-up CRL must not be published later than this date. Although RFC 5280 [3] proposes not to use CRLs that do not contain this date, some issuers do not include this field to avoid liability. The nextUpdate time in a CRL should not be earlier than the nextUpdate time in a previously issued CRL. This may be achieved by checking the nextUpdate time of the previously issued CRL or by using a constant period between thisUpdate and nextUpdate.

```
reasonCode ::= { CRLReason }

CRLReason ::= ENUMERATED {
  unspecified (0),
  keyCompromise (1),
  cACompromise (2),
  affiliationChanged (3),
  superseded (4),
  cessationOfOperation (5),
  certificateHold (6),
  removeFromCRL (8),
  privilegeWithdrawn (9),
  aACompromise (10) }
```

Listing 5.2 Reason code extension

revokedCertificates The `revokedCertificates` field contains the actual list of revoked certificates. If no certificate is revoked, then this list is not present. Revoked certificates are also called CRL entries. They consist of the following sub-fields.

userCertificate This field contains the serial number of the revoked certificate. In direct CRLs, this serial number and the issuer contained in the CRL issuer field determine the revoked certificate uniquely. Later in this section we explain how certificates can be identified in indirect CRLs.

revocationDate This entry contains the revocation date and time for the certificate in this entry. It is assigned by the CRL issuer and not by the certificate owner. The revocationDate of certificates that appear for the first time in this CRL must not be earlier than thisUpdate of the previous CRL. For such certificates, the time difference between the revocationDate and thisUpdate of the present CRL is called *revocation latency*. In addition to the two basic fields userCertificate and revocationDate, CRL entries may contain three extensions which are explained next.

reasonCode This extension describes the reason for the revocation of the certificate. Listing 5.2 shows the specification of the extension, which contains ten possible reasons. We explain the possible reasons.

unspecified There is no particular reason for the revocation. RFC 5280 proposes to not include the reasonCode extension instead of using this entry.

keyCompromise For end entity certificates only. The subject's private key or other security critical aspects regarding the subject are no longer secret or are suspected to be compromised.

cACompromise For CA certificates only. The subject's private key or other security critical aspects regarding the subject are no longer secret or are suspected to be compromised.

affiliationChanged The information on the subject is no longer correct. There is no indication of a compromise.

superseded There is a newer certificate replacing the one at hand. There is no indication of a compromise.

cessationOfOperation The certificate is not needed anymore. There is no indication of a compromise.

certificateHold The certificate is temporarily on hold and may be reactivated later.

removeFromCRL For use with delta CRLs only. The certificate is to be removed from the CRL as the certificate is expired or is no longer on hold.

privilegeWithdrawn A privilege contained in this certificate has been withdrawn.

aACompromise For attribute authority certificates only. The subject's private key or other security critical aspects regarding the subject are no longer secret or are suspected to be compromised.

invalidityDate This is the date when the event that makes revoking the certificate necessary happened. It may be earlier than the revocation time specified in revocationDate. For example, a key may be compromised some time before the compromise is discovered by the user. Clearly, it is not meaningful for this date to be later than the revocation date of the CRL entry.

certificateIssuer This extension contains the distinguished name of the issuer of the certificate revoked in this entry. Additionally, this entry may contain the issuer alternative name. If this extension is absent in the first entry of the CRL, then the issuer of this revoked certificate is the issuer of the CRL. If this entry is absent in any other entry of the CRL then the issuer of the revoked certificate is the issuer of the previous entry. In this extension the same encoding must be used as in the issuer distinguished name of the certificate.

5.2.2 CRL Extensions

Next, we describe possible CRL extensions. The extensions authority key identifier, issuer alternative name, and authority information access are the same as the corresponding certificate extensions. They are discussed in Sects. 2.3 and 9.6.1 respectively. In addition, there are a few more extensions.

CRL Number CRLs are issued sequentially. After an initial CRL has been issued, it is updated by issuing subsequent CRLs that contain newly revoked certificates and (at least) all the previously revoked certificates that have not expired. The CRL number allows organizing the CRLs by issuing time. The CRL number must not be smaller than 0. Typically, the CRL number of the first CRL is 0. Also, the CRL number of a CRL must be larger than the CRL numbers of all previously issued CRLs. If two CRLs happen to be issued at the same time by the same issuer, then their CRL numbers must be identical. The CRL number extension is mandatory.

```
IssuingDistributionPoint ::= SEQUENCE {
  distributionPoint            [0] DistributionPointName OPTIONAL,
  onlyContainsUserCerts        [1] BOOLEAN DEFAULT FALSE,
  onlyContainsCACerts          [2] BOOLEAN DEFAULT FALSE,
  onlySomeReasons              [3] ReasonFlags OPTIONAL,
  indirectCRL                  [4] BOOLEAN DEFAULT FALSE,
  onlyContainsAttributeCerts   [5] BOOLEAN DEFAULT FALSE }
```

Listing 5.3 Issuing distribution point extension

Delta CRL Indicator This extension indicates that the CRL is a delta CRL, which means that it contains only the certificates that have been revoked after a previous full CRL has been issued. If this extension is not present, then the CRL is a full CRL. The delta CRL Indicator extension must be marked critical because applications must be able to distinguish between full and delta CRLs.

The value of this extension is an integer. Using the delta CRL indicator, it is possible to construct a full CRL from this delta CRL and the previous full CRL. This is explained in more detail in Sect. 5.2.4.

Issuing Distribution Point The issuing distribution point extension allows specifying the location of the CRL. It also supports partitioning of CRLs by limiting the scope of the CRL. Listing 5.3 shows the specification of this extension.

The `distributionPoint` field specifies the location of this CRL. It has the format of CRL distribution points as explained in Sect. 5.3.1. If it is set, then it must contain at least one of the distribution points contained in the certificate. This enables a check that the location of this CRL matches the location where revocation information about a certificate is expected. If it is not set, then all unexpired certificates that are revoked within the scope of the CRL must be contained in it. This prevents further partitioning of CRLs within the same scope.

The extension also allows limiting the scope of the CRL to certain types. Possible types are certificates of end entities (`onlyContainsUserCerts` = true), certificates of certification authorities (`onlyContainsCACerts` = true), and attribute certificates (`onlyContainsAttributeCerts` = true). There must be no more than one type present in this extension. The extension may also limit the reason of revocation of the certificates contained in the CRL. If the `onlySomeReasons` field is present, then the CRL only contains certificates that are revoked for one of the reasons contained in this field.

If present, the issuing distribution point extension must be marked critical. The reason is that if an application is unable to evaluate this extension, then the status of certificates not contained in the present CRL must be considered unknown since the present CRL may only be a part of a partitioned CRL.

Freshest CRL This extension is sometimes also referred to as `delta CRL distribution point`. It is used when delta CRLs are issued to incrementally update full CRLs. It informs users where to locate the delta CRLs that are used

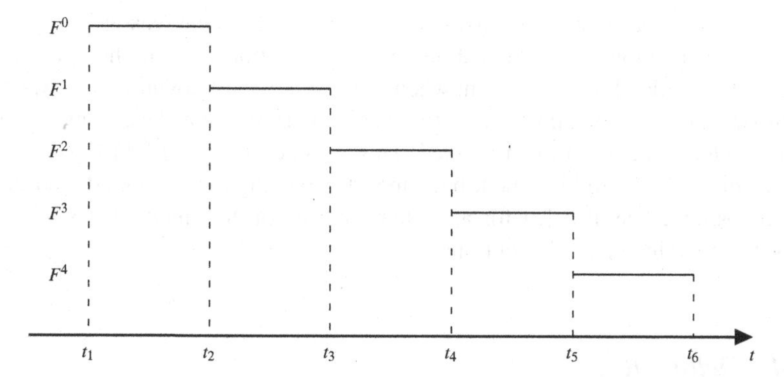

Fig. 5.2 CRLs issued at regular time intervals

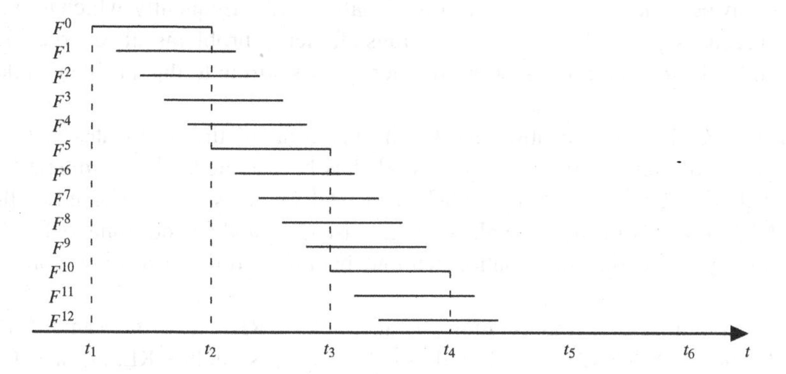

Fig. 5.3 Over-issued CRLs

in conjunction with the complete CRL. The extension has the form of a CRL distribution point extension as described in Sect. 5.3.1.

5.2.3 Issuing Time of a CRL

A very security-sensitive issue is the CRL issuing time. As we have seen, the issuing time of the CRL is stored in the field thisUpdate and the latest date at which the next CRL is published is written into the nextUpdate field. Many CRL issuers use a fixed time interval between thisUpdate and nextUpdate which can vary between a few hours or days. This is depicted in Fig. 5.2.

If the nextUpdate field is used, then a CRL must be issued not later than the time indicated in this field. This does not prohibit the CRL issuer from issuing CRLs before the nextUpdate time. Such CRLs are called over-issued CRLs. The use of those CRLs is depicted in Fig. 5.3.

There are various reasons for using over-issued CRLs. One reason is to provide immediate revocation information about individual certificates. Another reason is to make CRL downloads more efficient when a large number of downloads is expected. The thisUpdate and nextUpdate dates for over-issued CRLs may be chosen such that the time interval between them is fixed. This is done in $F^1 - F^4$ in Fig. 5.3. This requires many CRLs to be issued, but supports reducing the download load of the CRL repository. See also [2] for a further analysis on this method. As with full CRLs, the nextUpdate field is optional.

5.2.4 Delta CRLs

CRLs may become very large. Therefore, updating CRLs frequently, which is useful from a security point of view, causes serious efficiency problems since large CRLs have to be downloaded by the users frequently. A solution to this problem is delta CRLs.

A delta CRL is a revocation list which only contains the certificates that have been revoked after a certain complete CRL has been issued. This complete CRL is called *Base CRL* for the delta CRL. Delta CRLs are small compared to their base CRL. Users who have downloaded the Base CRL and an additional delta CRL have timely revocation information without having to download many complete CRLs.

We explain delta CRLs in more detail. Delta CRLs have the same ASN.1 specification as X.509v2 CRLs. To indicate that a CRL is a delta CRL, the delta CRL indicator (see Sect. 5.2.2) is used. This extension contains the CRL number of the Base CRL. A delta CRL can be combined with its base or any other complete CRL issued after the Base CRL, since such a CRL contains at least the same revocation information as the Base CRL.

In order to provide complete revocation information, a delta CRL must satisfy the following requirements with respect to its Base CRL or any complete CRL issued after the Base CRL.

- Complete CRL and delta CRL have the same issuer and the issuer uses the same signature key.
- Complete CRL and delta CRL have the same scope.
- The Base CRL number indicated in the delta CRL is smaller than or equal to the CRL number of the complete CRL.
- The CRL number of the complete CRL is smaller than the CRL number of the delta CRL.

There may be clients that are not able to process delta CRLs, for example if they do not understand the critical delta CRL indicator extension. For supporting these clients, CRL issuers can decide to issue a complete CRL with every delta CRL.

5.2.5 *Authority Revocation List*

Usually, CRLs contain certificates that belong to end entities and not to certification authorities. However, if it is necessary to revoke a CA certificate the issuer of the CRL may decide to insert it into a CRL that is used exclusively to revoke CA certificates. Such a revocation list is called an *authority revocation list (ARL).*

5.2.6 *Indirect CRLs*

Certificate issuers may decide to delegate revocation to a dedicated CRL issuer. For example, there may be several certificate issuers in a PKI. However, for efficiency and security reasons it is useful to have a single CRL issuer where the users can obtain the full revocation information. Another reason for delegating revocation is that the certificate issuer may not be able or willing to guarantee the required response times or is even offline.

A CRL issued by a dedicated CRL issuer who is not the certificate issuer is called *indirect CRL.* Issuers of indirect CRLs use a certificate with cRLSign as key usage. The CRL entries of an indirect CRL use the extension certificateIssuer (see Sect. 5.2.1) to inform the user about the issuer of the certificate. Moreover, indirect CRLs are specified as indirect by setting the appropriate flag in the issuing distribution point extension as discussed in Sect. 5.2.2. The use of delta CRLs based on indirect CRLs is permitted.

Unfortunately, not many applications are able to process indirect CRLs. Therefore, indirect CRLs are not widely used.

5.3 Certificate Extensions Related to Revocation

We discuss certificate extensions that are used to support revocation.

5.3.1 *CRL Distribution Points*

When an application uses a certificate, it must check whether or not this certificate has been revoked. When CRLs are used as a revocation mechanism, applications need information on how to locate and download the relevant CRLs. This information is contained in the *CRL distribution points* extension, which contains one or more distribution points (see Listing 5.4). Each distribution point contains a URI pointing to a location of a CRL. It also contains fields indicating which revocation reasons the CRL covers. The extension may also contain the issuer of the CRL, which is, for example, needed in indirect CRLs. For an example of such an extension see Sect. 9.1.3.

```
DistributionPoint ::= SEQUENCE {
  distributionPoint      [0]   DistributionPointName OPTIONAL,
  reasons                [1]   ReasonFlags OPTIONAL,
  cRLIssuer              [2]   GeneralNames OPTIONAL }

DistributionPointName ::= CHOICE {
  fullName                  [0]   GeneralNames,
  nameRelativeToCRLIssuer [1]   RelativeDistinguishedName }
```

Listing 5.4 CRL distribution point extension

5.4 OCSP

The advantage of CRLs is that applications can obtain the full revocation information at predictable points in time. This means that the applications do not have to be connected to the CRL issuer unless they wish to download a fresh CRL. But on the other hand, CRLs may become very large, downloading them becomes extremely time consuming, and storing may need a lot of space which may not be available, for example in mobile devices. Also, due to the potentially long time intervals between the publication of two subsequent lists, revocation information may not be up to date when it is used, in particular, shortly before nextUpdate.

This is why the *online certificate status protocol* (OCSP) was invented. It allows clients to query an OCSP server about the revocation status of individual certificates. The OCSP method has certain advantages over the CRL method. First, the revocation information has a chance to be always fresh. Users may obtain revocation information immediately after the certificate is revoked. However, some servers just query CRLs, which eliminates this advantage. Another advantage of OCSP is that it does not require much storage. Only the revocation information about the certificate under examination must be stored. On the other hand, in contrast to the CRL method, OCSP requires the applications that need revocation information to be online.

OCSP is specified in RFC 2560 [8]. Many PKIs implement OCSP and many clients are compliant with OCSP. As an example, Fig. 5.4 shows the configuration of an OCSP server in Firefox.

5.4.1 Functionality

We explain how OCSP works. OCSP clients send a request to an OCSP server about the revocation status of one or more certificates. Optionally, the requests may be digitally signed. For each certificate, the request submits the serial number of the certificate, the hash value of the issuer's DN, and the hash value of the issuer's public key. This information determines the certificates uniquely. Listing 5.5 shows the ASN.1 specification of an OCSP request.

The OCSP server replies to the request with three possible answers. If the status of more than one certificate is requested, then the answer contains information about

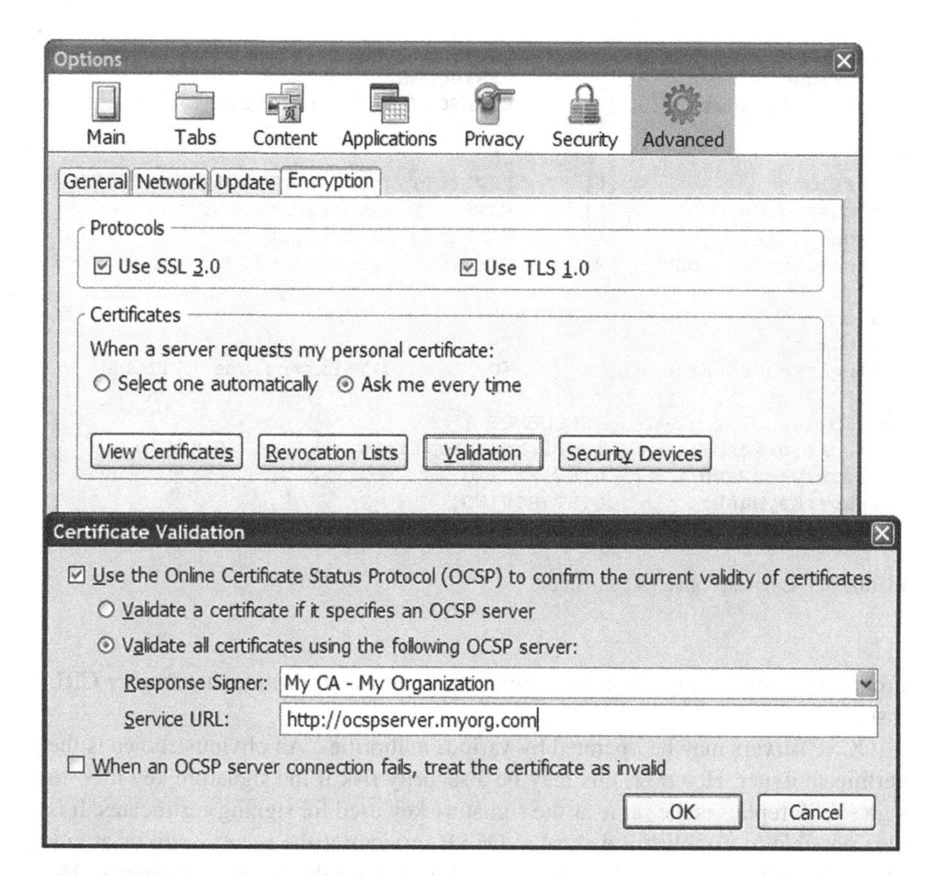

Fig. 5.4 Configuration of an OCSP server in Firefox

the status of all certificates. The first possible answer is *revoked*. This means that the certificate has been previously revoked. The second possible answer is *good*. It indicates that the certificate is not revoked. However, this answer does not mean that the certificate is still valid. It may have expired. Also, the answer "good" does not imply that the certificate exists. The third possible answer is *unknown*. This answer means that the certificate is not known to the OCSP server and that it is unable to give any answers about its status. If a protocol error has occurred, the OCSP server answers with an error message which is not signed. Details about OCSP responses are shown in Listing 5.6.

The meaning of `thisUpdate` and `nextUpdate` is analogous to the meaning of the corresponding elements of a CRL, i.e. thisUpdate indicates when revocation information has been updated by the OCSP server and nextUpdate refers to the latest date at which the revocation will be updated again. The time when the answer is signed is contained in the `producedAt` element. The reason for this information is that OCSP servers may prepare their answers in advance and use them later. If the

```
OCSPRequest        ::=      SEQUENCE {
  tbsRequest                       TBSRequest,
  optionalSignature      [0]       EXPLICIT Signature OPTIONAL }

TBSRequest         ::=      SEQUENCE {
  version                [0]       EXPLICIT Version DEFAULT v1,
  requestorName          [1]       EXPLICIT GeneralName OPTIONAL,
  requestList                      SEQUENCE OF Request,
  requestExtensions      [2]       EXPLICIT Extensions OPTIONAL }

Request            ::=      SEQUENCE {
  reqCert                          CertID,
  singleRequestExtensions          [0] EXPLICIT Extensions OPTIONAL }

CertID             ::=      SEQUENCE {
  hashAlgorithm          AlgorithmIdentifier,
  issuerNameHash         OCTET STRING,
  issuerKeyHash          OCTET STRING,
  serialNumber           CertificateSerialNumber }
```

Listing 5.5 Elements of an OCSP request

producedAt time is unacceptable, clients may choose to download a fresher CRL instead.

OCSP servers may be operated by various authorities. An obvious choice is the certificate issuer. However, this may be a security risk if the signature key used to sign OCSP replies is the same as the signature key used for signing certificates. It is also possible to establish a dedicated OCSP service (at the issuer's site or at any other site) with a distinct DN which is different from the certificate issuer's DN. This OCSP service must have a certificate of the issuer with the extended key usage extension set to the value `OCSPsigning`, which has the OID 1.3.6.1.5.5.7.3.9.

Applications that use OCSP must be able to decide whether the certificate of the OCSP server is revoked or not. It is possible to check the validity of the OCSP server certificate by inspecting a CRL. However, this reduces the efficiency advantage of OCSP. Another approach is to make the OCSP server certificate permanently valid by adding the extension `ocsp-nocheck` to it and choosing a short validity period for its certificate.

5.4.2 Extensions

It is possible for clients and servers to add extensions to their OCSP requests and responses. An overview over these extensions and their OIDs is presented in Table 5.1. This table also shows whether servers or clients can set the extensions. We explain the meaning of the extensions.

```
OCSPResponse ::= SEQUENCE {
  responseStatus         OCSPResponseStatus,
  responseBytes          [0] EXPLICIT ResponseBytes OPTIONAL }

ResponseBytes ::= SEQUENCE {
  responseType    OBJECT IDENTIFIER,
  response        OCTET STRING }

BasicOCSPResponse       ::= SEQUENCE {
  tbsResponseData       ResponseData,
  signatureAlgorithm    AlgorithmIdentifier,
  signature             BIT STRING,
  certs                 [0] EXPLICIT SEQUENCE OF Certificate
     OPTIONAL }

ResponseData ::= SEQUENCE {
  version               [0] EXPLICIT Version DEFAULT v1,
  responderID               ResponderID,
  producedAt                GeneralizedTime,
  responses                 SEQUENCE OF SingleResponse,
  responseExtensions    [1] EXPLICIT Extensions OPTIONAL }

SingleResponse ::= SEQUENCE {
  certID                  CertID,
  certStatus              CertStatus,
  thisUpdate              GeneralizedTime,
  nextUpdate          [0] EXPLICIT GeneralizedTime OPTIONAL,
  singleExtensions    [1] EXPLICIT Extensions OPTIONAL }

CertStatus ::= CHOICE {
  good         [0]    IMPLICIT NULL,
  revoked      [1]    IMPLICIT RevokedInfo,
  unknown      [2]    IMPLICIT UnknownInfo }
```

Listing 5.6 Elements of an OCSP response

Table 5.1 Extensions for use in OCSP

Name	OID	Included by
Nonce	1.3.6.1.5.5.7.48.1.2	Client & server
CRL references	1.3.6.1.5.5.7.48.1.3	Server
Acceptable response types	1.3.6.1.5.5.7.48.1.4	Client
Archive cutoff	1.3.6.1.5.5.7.48.1.6	Server
CRL entry extensions	2.5.29.21, 2.5.29.24, 2.5.29.29	Server
Service locator	1.3.6.1.5.5.7.48.1.7	Client

Nonce This extension is used to prevent replay attacks. The client sends a random value to the server. The server must include this value in its answer.

CRL References This extension points to the CRL where a revoked or suspended certificate can be found. This entry may contain the URL where the CRL can be found, the CRL number, or the issuing time of the CRL.

Acceptable Response Types There can be several types of OCSP answers. By using this extension in an OCSP request, clients indicate which types of response they are able to process. RFC 2560 [8] defines the *basic response type* for OCSP, which is expected to be supported by all servers and clients.

Archive Cutoff The *archive cutoff* is used to indicate whether a digital signature was valid when it was produced, even if the respective certificate is expired at the verification time. For that, OCSP responders need to retain revocation information on a certificate beyond the actual expiration date of the certificate. The *retention interval* indicates how long the OCSP server keeps revocation information after certificate expiration. After that interval, the OCSP responder does not have any information about the certificate's revocation history, and might declare revoked certificates as *good* (which is compliant to the definition of possible OCSP responses).

The archive cutoff is a value that indicates whether or not the OCSP responder still holds revocation information on the certificate in question. The actual value of the archive cutoff depends on the time the OCSP response is created, and therefore has to be calculated for each response. The value is obtained by subtracting the OCSP responder's retention interval from the date indicated by the producedAt element of the response.

The value of archive cutoff is a date that is used by the client to determine, via comparison with the certificate expiration date, the quality of the OCSP response. As long as the archive cutoff date is before the certificate expiration date, the OCSP response for a revoked certificate will be "revoked", even if the validation time is after the certificate expiration time. If the archive cutoff date is after the certificate expiration date, the OCSP response may be "good" although it has been revoked before the signature in question was generated.

CRL Entry Extensions All extensions of a CRL entry can also be included in the OCSP requests and answers. For example, the server may provide the invalidity date of a certificate or the reason for the revocation. See Sect. 5.2.1 for a description of these extensions.

Service Locator OCSP servers may operate as proxy servers for other OCSP servers. In this extension of the OCSP request, the OCSP server is informed about the location of the OCSP server which is responsible for answering OCSP requests concerning the certificate in the request.

5.4.3 Lightweight OCSP

OCSP servers that are responsible for answering requests for many certificates may run into efficiency problems in regard to bandwidth and computing power. This is even more likely in mobile application environments where bandwidth and computing power is limited. This is why a lightweight variant of the OCSP protocol was proposed in [4].

This OCSP variant reduces the options for OCSP requests in order for them to be evaluated much more efficiently. For example, requests must not ask about more than one certificate, they must use SHA1 to hash public key and issuer name and must not use extensions. Responses must contain the basic OCSP response and it is recommended that they contain only one response and no extensions.

5.4.4 Design of an OCSP Server

When designing an OCSP server, it must be decided how the server obtains revocation information. One possibility for the OCSP server is to extract the revocation information from a CRL. If a certificate in a request is contained in the CRL, then the OCSP server answers "revoked". When using this strategy, the OCSP information is not fresher then the CRL information, which eliminates one of the advantages of OCSP: fresher revocation information. However, the OCSP revocation information is still much smaller than the CRL revocation information. An alternative approach is to send revocation information concerning individual certificates directly to the OCSP server. In this case, the OCSP server answers "revoked" whenever it has such revocation information. This solution offers much better timeliness than using CRLs.

If the OCSP server has no revocation information that justifies the answer "revoked", it must choose between "good" and "unknown". Again, different policies are possible. One possibility is that the OCSP server answers "good" if it has access to revocation information from the issuer of the certificate in the request and "unknown" otherwise. However, some OCSP servers are more restrictive. They only answer "good" if they are convinced of the existence of the certificate contained in the request. For example, they may have access to a database containing all existing certificates.

5.5 Other Revocation Mechanisms

5.5.1 Novomodo

A special method of revocation was proposed in [6]. This method is called *Novomodo* and works as follows. With each certificate they create, CAs generate two random strings X_0 and Y_0 which remain secret. In addition to their standard content, certificates have two additional fields. The first field contains the hash value Y_1 of Y_0 and the second field contains the hash value X_{365} obtained by hashing X_0 365 times.

As long as a certificate is not revoked, the CA publishes the value X_{365-i} obtained by hashing X_0 $365-i$ times on the ith day of the certificate's validity. Applications that wish to evaluate the revocation status of a certificate on the ith day of its validity, hash the value X_{365-i} i times. If the result matches X_{365}, then the certificate is valid. In order to revoke a certificate, the CA publishes Y_0. Then all applications can verify for themselves that the certificate has been revoked by comparing the hash value of Y_0 to Y_1.

The authenticity of Y_1 and X_{365} is protected by the signature of the certificate. An optimization of this scheme uses hash trees. It reduces the number of hash value evaluations. It has been presented in [5].

5.5.2 *Short-Lived Certificates*

Revocation can be avoided if the validity of the certificate is very short, for example, 1 day. Even if revocation turns out to be necessary during the short validity period of the certificate, revoking it is of no use since the certificate expires very soon anyway and all revocation methods have a considerable revocation delay. In [7], short-lived certificates are discussed and analyzed. A special variant of this method is proposed in [9], where certificates are issued on demand. For example, CV certificates are short-lived certificates.

5.6 Revocation in PGP

There is no central revocation mechanism such as CRLs in PGP. Instead, owners of public keys can revoke their keys themselves by creating a signed statement which asserts that this key is revoked. This statement is sometimes called a *revocation certificate*. The revocation certificate is signed using the standard user signature key. As with the normal certification of the user ID and public key, a signature is calculated over this data (see Sect. 2.6). This signature however is of another type than the certification signature. It has another value in the type octet of the signature and therefore this certificate can be recognized as a revocation.

For revoking a PGP certificate, the user may send it directly to communication partners or upload it to the key servers used in PGP. The other users then need to import the revocation certificate to their key ring to mark the certificate as revoked.

Typically, the PGP revocation certificate is created immediately after the key generation. This is because the key that is needed to sign the revocation certificate may not be available at the time of the actual revocation. For example, a user may

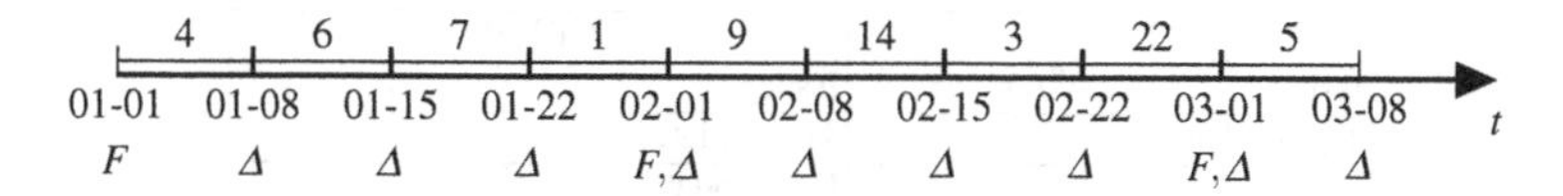

Fig. 5.5 Revocation in the course of time

Table 5.2 CRL numbers and Base CRLs

CRL	CRL number	Base CRL number
Complete CRL at 01-01		
Delta CRL at 01-15		
Complete CRL at 02-01		
Delta CRL at 02-01		
Delta CRL at 02-08		
Delta CRL at 03-08		

have forgotten the password required for unlocking the key or may have lost the key. The revocation certificate must be protected in order to prevent unauthorized users from revoking the key.

5.7 Exercises

5.1. Check whether a user has all the revocation information contained in the full CRL F^{14} when one of the following sets of full CRLs F and delta CRLs Δ is given. Here, the delta CRL j with Base CRL i is denoted by Δ_i^j.

1. $\{F^1, \Delta_1^5, \Delta_5^{10}, \Delta_{10}^{12}, \Delta_{10}^{15}\}$
2. $\{F^1, \Delta_1^5, F^5, F^{12}, \Delta_{10}^{14}\}$
3. $\{F^1, \Delta_2^5, \Delta_5^{10}, \Delta_{12}^{14}, \Delta_{10}^{15}\}$

5.2. A CA starts its operation on 1 December 2011. As revocation mechanism the CA uses CRLs whose scope is all certificates. The CA issues one complete CRL at the beginning of each month. The first complete CRL is issued on 1 January 2012. It contains 12 revoked certificates. In addition, the CA issues delta CRLs on the 1st, on the 8th, on the 15th, and on the 22nd of each month. The Base CRL for the delta CRLs is the freshest complete CRL that was issued previously. For example, the Base CRL of the delta CRL issued on 1 February 2012 is the full CRL issued on 1 January 2012.

Figure 5.5 shows several CRL issuing dates in 2012 and revoked certificates. For example, between 8 January and 15 January, six certificates were revoked.

1. What is the CRL number ($\in [0, 9]$) and the Base CRL number of the CRLs given in Table 5.2. Select the numbers such that the numbering of the CRLs is consistent.
2. Which complete CRL has the most CRL entries? How many?

CRL A	
Issuer:	CA1
ThisUpdate:	
NextUpdate:	
Revoked Certificates:	
Serial No.: ...	
X509v2 CRL Extensions:	
CRL Number:	non-critical 143

CRL B	
Issuer:	CA1
ThisUpdate:	
NextUpdate:	
Revoked Certificates:	
Serial No.:	64
Serial No.:	128
X509v2 CRL Extensions:	
CRL Number:	non-critical 143
Delta CRL Indicator: Base CRL Number:	critical 138

Fig. 5.6 CRLs A and B with missing ThisUpdate and NextUpdate values

CRL C	
Issuer:	CA1
ThisUpdate:	
NextUpdate:	
Revoked Certificates:	
Serial No.:	32
Serial No.:	16
X509v2 CRL Extensions:	
CRL Number:	non-critical 138

Fig. 5.7 CRL C with missing ThisUpdate and NextUpdate values

CRL D	
Issuer:	CA1
ThisUpdate:	2011-10-15
NextUpdate:	2011-11-01
Revoked Certificates:	
Serial No.:	456
X509v2 CRL Extensions:	
CRL Number:	non-critical
Delta CRL Indicator: Base CRL Number:	critical

CRL E	
Issuer:	CA1
ThisUpdate:	2011-10-15
NextUpdate:	2011-11-15
Revoked Certificates:	
Serial No.:	232
Serial No.:	136
Serial No.:	164
Serial No.:	987
Serial No.:	456
X509v2 CRL Extensions:	
CRL Number:	non-critical

Fig. 5.8 CRLs with missing CRL number and Base CRL number values

3. Which delta CRL has the most CRL entries? How many?
4. A user has downloaded all delta CRLs. Which complete CRLs does this client need in order to have complete revocation information on 3 March?

5.3. The issuer with DN "CN = CA1" publishes revocation information in the form of CRLs and delta CRLs (for the scope "all certificates that this CA issues").

1. Enter the thisUpdate and nextUpdate values of the CRLs A, B, and C (shown in Figs. 5.6 and 5.7) that allow for consistency among the CRLs if no two complete

Fig. 5.9 An indirect CRL

CRL A	
Issuer:	CN=First CA, C=DE
ThisUpdate:	2011-04-25
NextUpdate:	2011-05-25
Revoked Certificates:	
Serial No.:	1
Serial No.:	2
Serial No.:	3
Certificate Issuer:	CN=Second CA, C=DE
Serial No.:	4
Serial No.:	5
Serial No.:	6
Certificate Issuer:	CN=Third CA, C=DE
Serial No.:	7
Certificate Issuer:	CN=Forth CA, C=DE
X509v2 CRL Extensions:	
CRL Number:	non-critical 500

Table 5.3 Revoked certificates

Issuer	Serial number	Revoked
CN = First CA, C = DE	1	
CN = First CA, C = DE	2	
CN = First CA, C = DE	3	
CN = First CA, C = DE	4	
CN = First CA, C = DE	5	
CN = Second CA, C = DE	1	
CN = Second CA, C = DE	2	
CN = Second CA, C = DE	3	
CN = Second CA, C = DE	4	
CN = Second CA, C = DE	5	
CN = Second CA, C = DE	6	
CN = Third CA, C = DE	6	
CN = Third CA, C = DE	7	
CN = Forth CA, C = DE	6	
CN = Forth CA, C = DE	7	

CRLs are issued on the same date and the permitted dates are 2011-08-10, 2011-08-15, and 2011-09-10. In addition, enter the serial numbers of the revoked certificates of CRL A.

2. Can a delta CRL contain more certificates than its Base CRL? Give a reason for your answer.
3. Consider the CRLs D and E shown in Fig. 5.8. Choose appropriate CRL Number and Base CRL Number values from $\{234, 321, 333\}$.

5.4. Consider the indirect CRL shown in Fig. 5.9. Enter the values "true" or "false" in the column "revoked" of Table 5.3 that indicate whether a certificate is revoked or not.

References

1. A. Årnes, M. Just, S.V. Knapskog, S. Lloyd, H. Meijer, Selecting revocation solutions for PKI, in *Proceedings of NORDSEC 2000 Fifth Nordic Workshop on Secure IT Systems*, 2000. http://www.pvv.ntnu.no/~andrearn/certrev/
2. D.A. Cooper, A model of certificate revocation, in *Proceedings of the 15th Annual Computer Security Applications Conference (ACSAC'99)*, Scottsdale, 1999, pp. 256–264
3. D. Cooper, S. Santesson, S. Farrell, S. Boeyen, R. Housley, W. Polk, Internet X.509 public key infrastructure certificate and certificate revocation list (CRL) profile, in *IETF Request for Comments*, 5280, May 2008
4. A. Deacon, R. Hurst, The lightweight online certificate status protocol (OCSP) profile for high-volume environments, in *IETF Request for Comments*, 5019, Sept 2007
5. F.F. Elwailly, C. Gentry, Z. Ramzan, QuasiModo: efficient certificate validation and revocation, in *Proceedings of the 7th International Workshop on Theory and Practice in Public Key Cryptography, PKC 2004*, Singapore, 2004. Volume 2947 of Lecture Notes in Computer Science, pp. 375–388
6. S. Micali, Novomodo – scalable certificate validation and simplified PKI management, in *Online Proceedings of the 1st Annual PKI Research Workshop*, 2002. http://www.cs.dartmouth.edu/~pki02/
7. M. Myers, Revocation: options and challenges, in *Proceedings of Financial Cryptography, Second International Conference, FC'98*, Anguilla, 1998. Volume 1465 of Lecture Notes in Computer Science, pp. 165–171
8. M. Myers, R. Ankney, A. Malpani, S. Galperin, C. Adams, X.509 Internet public key infrastructure online certificate status protocol – OCSP, in *IETF Request for Comments*, 2560, June 1999
9. K. Scheibelhofer, PKI without revocation checking, in *Online Proceedings of the 4th Annual PKI R&D Workshop*, Apr 2005. http://middleware.internet2.edu/pki05/proceedings/

Chapter 6
Validity Models

In this chapter we deal with validity models for digital signatures in the hierarchical trust model. In order to explain what we mean by this, we start with an example. Paul sells his house to Anna on 1 October 2009. Paul signs the sales contract digitally. The certificate that authenticates Paul's signature verification key expires on 31 July 2010. Should Paul's signature still be considered valid after the certificate has expired? In the most common validity models, for example in the *shell model* from the PKIX standard, the answer is "no". This does not seem to make much sense since the transaction is still valid. Therefore, the German signature law requires an electronic signature to be valid independent of the expiration of the certificate as long as the certificate was valid at the time the signature was created. This so-called *chain model* is a completely different validity model for signatures. In this chapter we discuss the various validity models for digital signatures in more detail.

6.1 The Shell Model

In this section we describe the shell model. It is proposed in RFC 5280 (Sect. 6 of [1]) and used in the certification path validation algorithm described in Sect. 9.4.1. The model is shown in Fig. 6.1.

We explain under what conditions a signature is valid in the shell model. Suppose that Alice has signed a document. Bob wishes to verify this signature. Denote by $C_0, C_1, \dots, C_n$, $n \in \mathbb{N}$, the certification path that is used by Bob to verify the authenticity of Alice's verification key. Here, C_0 is a certificate for the corresponding verification key. In the shell model, Alice's signature is valid if and only if at verification time all certificates in the certification path $C_0, \dots, C_n$ are valid. This means that in the shell model the signature becomes invalid as soon as any of the certificates in the path expires or is revoked.

Example 6.1. Consider the three certificates shown in Fig. 6.2:

Suppose Alice signed a document on 2011-06-15. According to the shell model, this signature is valid on 2011-08-01 because all certificates in the chain are valid.

J.A. Buchmann et al., *Introduction to Public Key Infrastructures*,
DOI 10.1007/978-3-642-40657-7_6, © Springer-Verlag Berlin Heidelberg 2013

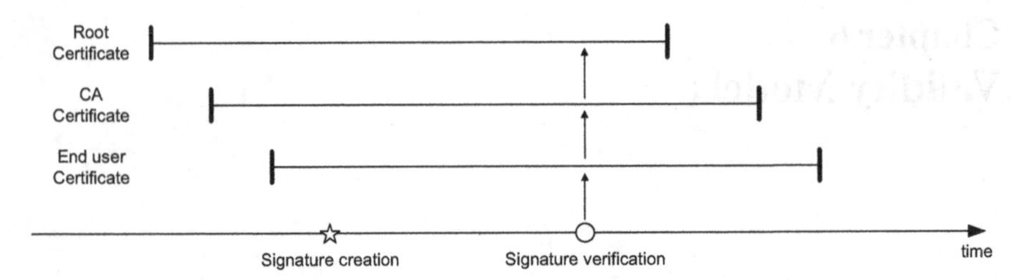

Fig. 6.1 The shell model

Certificate 1	
Serial No.:	1
Issuer:	CN=Root-CA
NotBefore:	2011-01-01
NotAfter:	2011-12-31
Subject:	CN=Root-CA
Public Key:	key-0x654121
X509v3Extensions:	
Subject Key Identifier: keyId: 12:AB:45:76:F8:98	
Authority Key Identifier: keyId: 12:AB:45:76:F8:98	

Certificate 2	
Serial No.:	12
Issuer:	CN=Root-CA
NotBefore:	2011-03-01
NotAfter:	2012-02-28
Subject:	CN=Sub-CA
Public Key:	key-0x943566
X509v3Extensions:	
Subject Key Identifier: keyId: CA:53:91:2A:E6:22	
Authority Key Identifier: keyId: 12:AB:45:76:F8:98	

Certificate 3	
Serial No.:	12
Issuer:	CN=Sub-CA
NotBefore:	2011-05-01
NotAfter:	2012-04-30
Subject:	CN=Alice
Public Key:	key-0x567812
X509v3Extensions:	
Subject Key Identifier: keyId: 2D:B8:8F:2F:64:48	
Authority Key Identifier: keyId: CA:53:91:2A:E6:22	

Fig. 6.2 Certificates in the shell model

The signature is invalid on 2012-02-01 because Certificate 1 has expired. Likewise, the signature is invalid on 2012-06-05 because Certificates 1 and 2 have expired. Figure 6.3 illustrates this example.

Today, the shell model is the one used everywhere on the Internet. This is appropriate in all applications where signing and verification times are very close to each other. Examples of such applications are challenge-response authentication mechanisms or email authentication. However, for contract signing this model appears to be inappropriate. Signatures that prove the authenticity of a contract

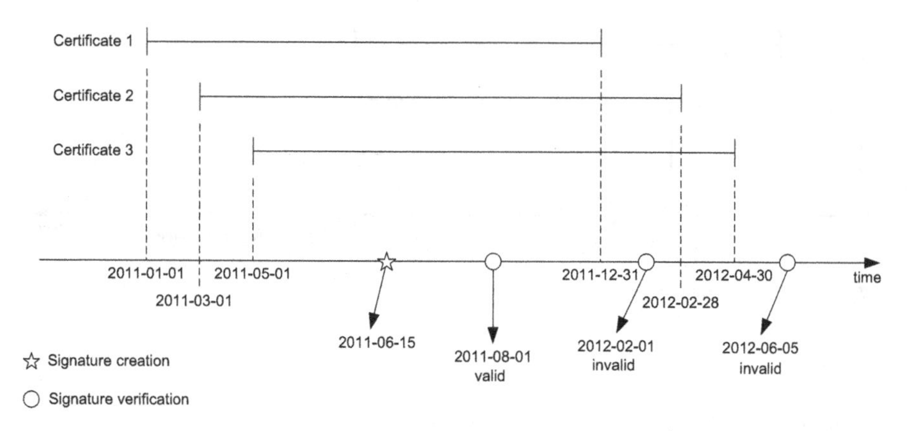

Fig. 6.3 Verification of a signature according to the shell model

are expected to prove this authenticity much longer than the validity period of the certificates in the corresponding verification path. This is why the chain model, which is explained in the next section, has been introduced.

6.2 The Chain Model

We now explain the chain model (Fig. 6.4).

As in the last section, suppose that Alice has signed a document, Bob wishes to verify this signature, and $C_0, C_1, \ldots, C_n$, $n \in \mathbb{N}$ is the certification path that is used by Bob to verify the authenticity of Alice's verification key. In the chain model, Alice's signature is valid if and only if all certificates in the corresponding chain were valid when the private keys corresponding to the keys that they certify were used for signing. More precisely, C_i certifies the verification key for the signature on C_{i-1}, $1 \leq i \leq n$. The certificate C_i must be valid when C_{i-1} is signed. In addition, C_0 is the certificate for the verification key for Alice's signature. Also this certificate must be valid when Alice issues her signature. This means that in contrast to the shell model, in the chain model the validity of a signature is independent of the verification time for this signature.

Example 6.2. Consider the three certificates from Example 6.1 and Fig. 6.5:

Any document that Alice signs between 2011-05-01 and 2012-04-30 is valid in the chain model because her certificate (Certificate 3) is valid at signature time. The signature on Certificate 3 is valid because at its issuing time Certificate 2 was valid. Likewise, Certificate 2 was valid because at its issuing time Certificate 1 was valid.

Therefore, the signature created on 2011-06-15 is valid both on 2012-02-01 and 2012-06-05. However, the signature created on 2012-05-12 is invalid because Alice's certificate is expired.

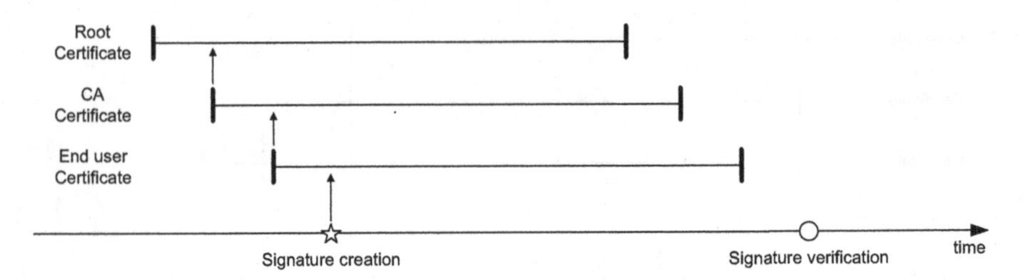

Fig. 6.4 The chain model

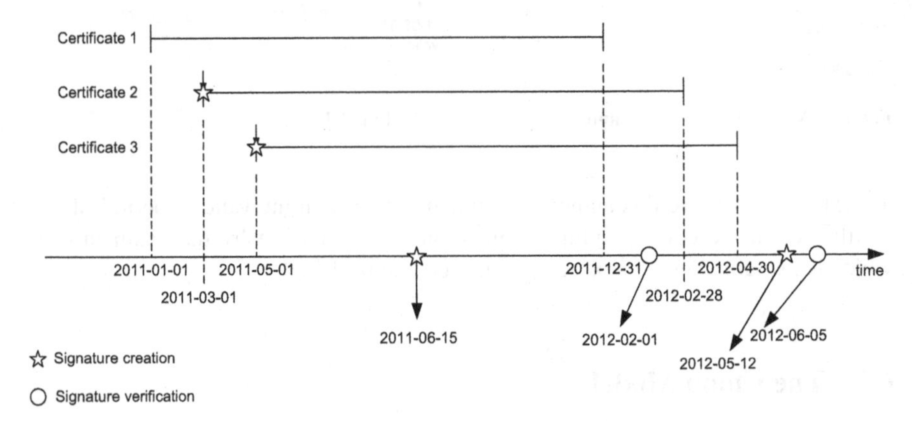

Fig. 6.5 Verification of a signature according to the chain model

The chain model is used in Germany for verifying legally binding electronic signatures because such signatures may be used for contract signing.

The chain model supports long validity periods for digital signatures. However, it has certain drawbacks. If Alice issues a signature and later a certificate in the chain that certifies Alice's verification key is revoked, the signature remains valid. This may have serious effects if the revocation reason is key compromise.

6.3 The Modified Shell Model

We now explain the *modified shell model*, which is also called *hybrid model*. It is a mixture of the chain and the shell models and is implicitly mentioned in RFC 5126 [2].

In the modified shell model, which is shown in Fig. 6.6, an end user signature is valid if it is valid in the shell model at creation time. This means that at creation time of the end user signature, all certificates in the relevant certification path must be valid. Thus, the validity of an end user signature is independent of the verification

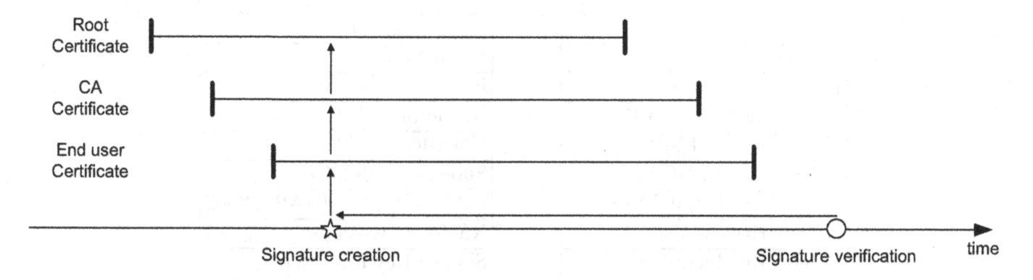

Fig. 6.6 The modified shell model aka hybrid model

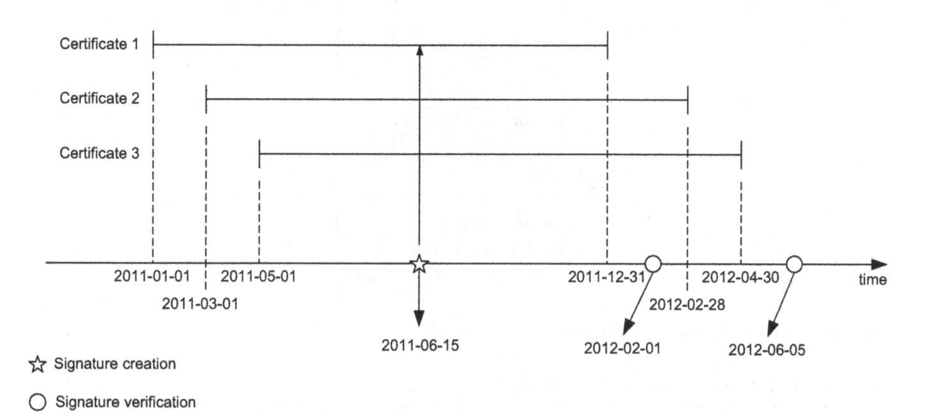

Fig. 6.7 Verification of a signature according to the modified shell model

time for that signature. This allows such a signature to be valid for a very long time. However, in contrast to the chain model, the validity of the end user signature is independent of the creation time of certificate signatures. This makes signature verification much simpler. Also, this implies that none of the certificates in the relevant certification path must be expired or revoked at the time of the creation of the end user signature. However, expiration or revocation of one of the certificates in the certification path after signature generation has no influence on the signature validity. As in the chain model, this may be problematic, for example, if such a certificate has been revoked because of key compromise.

Example 6.3. Consider the three certificates from Example 6.1 and Fig. 6.7.

Any document that Alice signs between 2011-05-01 and 2011-12-31 is valid in the modified shell model because all three certificates in the certification path are valid at signature time.

One application of the modified shell model is advanced electronic signatures as specified in RFC 5126 [2]. These signatures are used for long-term scenarios. This specification proposes time-stamping a digitally signed document directly after it is digitally signed—as a proof of existence of the document—including its signature at

Certificate A	
Serial No.:	1
Issuer:	RootCA
NotBefore:	2010-01-01
NotAfter:	2012-05-31
Subject:	RootCA
Public Key:	key-0x75590812
X509v3Extensions:	
Subject Key Identifier: keyId: 0A:32:F0:35:5F:3F	
Authority Key Identifier: keyId: 0A:32:F0:35:5F:3F	

Certificate B	
Serial No.:	3
Issuer:	RootCA
NotBefore:	2010-03-01
NotAfter:	2012-03-31
Subject:	SubCA
Public Key:	key-0xAB328602
X509v3Extensions:	
Subject Key Identifier: keyId: 60:43:6C:E5:9C:88	
Authority Key Identifier: keyId: 0A:32:F0:35:5F:3F	

Certificate C	
Serial No.:	5
Issuer:	SubCA
NotBefore:	2010-05-01
NotAfter:	2012-07-31
Subject:	Alice
Public Key:	key-0x11FE3665
X509v3Extensions:	
Authority Key Identifier: keyId: 60:43:6C:E5:9C:88	

Fig. 6.8 Certificates used in different validity models

Table 6.1 Validation in different models

Signature creation time	Shell model	Modified shell model	Chain model
2010-04-20			
2011-03-15			

signature generation time. Moreover, the specification proposes time-stamping the relevant certification path and the corresponding revocation information. Another application of the modified shell model is described in [3].

6.4 Exercises

6.1. In Fig. 6.8, three certificates are given.

1. Alice signs a document on 2010-04-20 and 2011-03-15. Enter the result of the signature validation on 2012-01-01 into Table 6.1. Possible values are valid or invalid.
2. Alice signs a document on 2011-03-15, 2012-05-24, 2012-07-08, and 2012-10-19. Enter the result of the signature validation on 2013-01-01 into Table 6.2. Possible values are valid or invalid.
3. Specify the time period in which Alice can create a valid signature for the different validity models. Enter the values into Table 6.3.

Table 6.2 Validation for different signature times

Signature creation time	Shell model	Modified shell model	Chain model
2011-03-15			
2012-05-24			
2012-07-08			
2012-10-19			

Table 6.3 Valid signature creation times

Model	From	Until
Shell		
Modified shell		
Chain		

References

1. D. Cooper, S. Santesson, S. Farrell, S. Boeyen, R. Housley, W. Polk, Internet X.509 public key infrastructure certificate and certificate revocation list (CRL) profile, in *IETF Request for Comments*, 5280, May 2008
2. D. Pinkas, N. Pope, J. Ross, CMS advanced electronic signatures (CAdES), in *IETF Request for Comments*, 5126, Feb 2008
3. U. Resnitzky, The directory-enabled PKI appliance: digital signatures made simple, approach and real world experience, in *6th Annual PKI R&D Workshop*, Gaithersburg, Apr 2007. http://middleware.internet2.edu/pki07/proceedings/

Chapter 7
Certification Service Provider

In Chap. 3 we explain the hierarchical trust model for PKIs. In this trust model, certification authorities that issue certificates play an important role. However, issuing certificates is not sufficient. The certificates must be maintained and additional information must be provided during the entire life cycle. The entity that is responsible for certificate life cycle management is called the *certification service provider* (CSP). A CA is only one component of a CSP. Another component is the *registration authority* (RA), which registers certificate applicants and collects all information relevant for issuing certificates. Other possible components of a CSP are, for example, a *directory service*, which publishes information concerning certificates, and a *revocation service*, which issues revocation information. Sometimes, certification service providers are also called *trust centers*. In this chapter, we start by explaining the life cycle of a certificate. Then, we describe the CSP components. In the next chapter, we discuss certificate policies that govern the operation of CSPs in a PKI.

7.1 Certificate Life Cycle

Certificate life cycles may vary according to the organization of a PKI and the certificate usage. Here, we describe a typical certificate life cycle. It can be divided into three phases.

7.1.1 Certificate Generation Phase

The certificate life cycle starts with a *certificate application* by some entity. This entity may be the intended owner of the certificate or some other authorized entity, for example, someone who is responsible for IT security in some institution of which the intended certificate owner is an employee. The application is followed by the

J.A. Buchmann et al., *Introduction to Public Key Infrastructures*,
DOI 10.1007/978-3-642-40657-7_7, © Springer-Verlag Berlin Heidelberg 2013

registration of the intended certificate holder. In the registration, the information relevant for issuing the certificate is collected and verified. This information typically includes the name of the intended certificate owner and some contact information such as an email address. It may also include the public key to be contained in the certificate if the corresponding key pair is generated by the intended user. After application and registration, the certificate is *issued*. If the key pair is not generated by the certificate applicant, the key pair is generated by a trusted third party such as the card manufacturer or the CSP while issuing the certificate. The CSP creates the certificate and signs it digitally. The CSP may also back up the private key corresponding to the public key in the certificate. For example, this is useful for decryption keys. For such keys, a backup prevents encrypted data from becoming inaccessible in the case where the private key owner loses the private decryption key. However, key backup opens security risks since the CSP has knowledge of private keys and with this knowledge the power to decrypt information of its users. Once a certificate is issued, it is *delivered* to and *accepted* by its owner. Without owner acceptance the certificate must not be used since it may contain incorrect information. The final step of the generation phase is *publication* of the certificate since it is intended to be used by many entities for establishing trust in the public key in the certificate belonging to the certificate owner.

7.1.2 Certificate Validity Phase

The main purpose of a certificate is to prove that the public key contained in the certificate belongs to the certificate owner. To this end, the certificate is retrieved from some directory. This may be a local directory of the user or a directory operated by some directory service. The certificate user validates the certificate, which includes verifying the certificate signature and checking the validity period of the certificate and its revocation status. But certificate validation may be much more involved. For example, if the certificate is an element of a certificate chain, the validity of the full chain must be checked. This is explained in Chap. 9.

When a certificate expires, the subject in the certificate may want to continue using the public key, for example, if he or she wants the smart card containing the corresponding private key to remain usable. In such a case, the certificate may be reissued before it expires. The re-issued certificate has the same content as the original certificate and a new expiration date.

7.1.3 Certificate Invalidity Phase

Typically, certificates have a limited validity period. For example, the validity of an X.509 certificate ends on the notAfter date in the certificate. Another way of invalidating certificates is to revoke them before they expire. Revocation is discussed in Chap. 5.

When the certificate of an encryption key is invalid, the corresponding public key must not be used anymore for encryption. Likewise, private keys corresponding to public keys of expired certificates must not be used to generate signatures anymore. However, frequently, the certificate cannot simply be deleted. In the chain and modified shell signature validity models, expired certificates may still be required for authenticating signature verification keys. Also, in long-term authentication protocols, even revoked certificates may have to be used. Therefore, invalid certificates may have to be archived.

7.2 Registration Authority

In the certificate generation phase, registration is a very important process in which the data necessary for generating certificates are collected and validated. This process is taken care of by a dedicated component of the CSP, the registration authority. Reasons for operating a separate registration authority may be found in App. A of [2]. For example, these are security reasons. Registration requires a lot of interaction with PKI users, while certificate issuance is the task of a few operators and must be protected from interference by unauthorized persons.

Registration is a complex process which may be implemented in various ways depending on operational and security requirements. We explain important steps of the registration process.

Registration starts with *establishing the identity* of the intended certificate owner. If the intended owner is a person, then the following information may be collected: first and last name, citizenship, place and date of birth, email address, employer, and biometric data such as fingerprints. This is more information than actually will be contained in the certificate. It serves several other purposes. For example, it allows the registration authority to decide whether the intended certificate owner is eligible to apply for a certificate. This data may be collected online or the certificate applicant may be required to register in person and to present some proof of identity such as an identity card. The data may also be retrieved from existing databases. In addition, the registration authority collects *contact information* and *billing data* of the certificate applicant. This may include postal address and telephone number. The information that the RA collects may become important in future legal disputes. If the certificate owner is intended to be a computing device, analogous information is gathered.

During registration, the *preferences* of the certificate holder are collected. Examples of such preferences are the cryptosystem to be used with the certified public key, parameters such as the key length, the delivery and billing method for the certificate, and possibly a pseudonym.

If the public key that is to be certified and the corresponding private key are generated by the certificate holder, the certificate applicant delivers the public key to the CSP as part of the registration process. Common delivery protocols and formats are PKCS#10, XKMS, and CRMF. PKCS#10 is most commonly used.

See also Sect. 7.5 for details on these protocols. Before the registration authority accepts a public key, the applicant must provide a *proof of possession* (PoP) of the corresponding private key. This prevents the CSP from issuing wrong certificates. Each protocol deals with PoP differently. For example, PCKS#10 implements PoP by requiring the applicant of a certificate to sign a PKCS#10 request. This means that he or she is in possession of the private key. This assumes that the private key can be used for signing even if it is a decryption key. CRMF provides three mechanisms to implement PoP. One is to reveal the private key to the CSP. A second is to exchange challenge-response messages between the CSP and the certificate user. For example, the CSP encrypts a random value and expects the user to return this value or the user sends a signed request. A third method is to create the certificate, encrypt it and let the user decrypt it.

During registration it may also be important to establish the possibility for a secure channel between the certificate owner and the CSP that is independent of the key in the certificate. For example, this may be required when the certificate owner wishes to revoke the certificate. A possible implementation of such a secure channel may be based on a secret password that is agreed upon in the registration phase. Upon collecting the registration information, the registration authority archives it for later reference. The data must be protected according to applicable regulations.

Next, the registration authority verifies that the entity applying for a certificate with the properties specified in the applicant's preferences is authorized to receive such a certificate. For example, authorization may depend on the age of an applicant or his or her position in an institution. If issuing a certificate is compliant with applicable rules, the registration authority creates a *unique digital name*. This digital name will be contained in the certificate and is assigned to not more than one entity. It must not be used in a certificate which belongs to another entity. Registration authorities may apply certain rules or restrictions when creating unique digital names. Such rules help to make digital names meaningful. For example, if the certificate belongs to an SSL server, then it makes sense to use the server domain name as digital name. In addition, the name should support other functions such as searching for certificates of a certain entity.

Since the registration process frequently requires applicants to appear in person, the registration authority may operate many regional registration offices which are sometimes referred to as *local registration authorities* (LRAs). This is particularly useful for international organizations with branches in many countries. To make certificate registration even more smooth it may be embedded into some other registration process, for example, of the human resources department of some organization.

If the registration has been completed successfully, the registration authority submits a certificate request to the certification authority. This request contains all information which is required by the CA to issue a certificate.

7.3 Certification Authority

Upon receiving certification requests from the registration authority, the certification authority issues a certificate. The CA creates the certificate in accordance with the applicable standard, for example X.509, and signs it using its secret signature key.

If the certification request does not contain a public key, the CA generates a key pair on behalf of the future certificate owner. The CA includes the public key of that pair in the certificate and delivers the private key to the certificate owner. Typically, the CA stores the private key within a personal security environment (PSE) such as a smart card. The use of the smart card may be protected by the PIN that has been selected by the user in the registration process. Alternatively, the CA may also create a PIN for the user that is delivered separately from the PSE containing the private key. The CA may also use alternative methods for delivering the private key. The PSE is delivered to the user usually by mail or is picked up by the user directly at the CSP. Software PSEs may be delivered by email or downloaded from the CSP. They may also be directly installed at the client side by the system administration. In any case, they must be protected appropriately.

In addition to issuing certificates, the CA may also provide revocation information. If certificate revocation lists are used, the CA may generate and sign those lists. If the OCSP protocol is applied, the CA may sign the OCSP responses.

There are many ways for the CSP to receive revocation requests. One possibility is to establish a hotline which the certificate users can reach. By providing the certificate data as well as some authentication data such as a revocation password, the users can revoke their certificates. Providing the revocation password is necessary to avoid unauthorized revocations. Another possibility is to use a Web page instead of a hotline. It also uses a revocation password. The CSP may also initiate a revocation request. This is practiced when a certificate user is not eligible to use the certificate anymore.

Another task of the CA is to back up the private key that it has created for the certificate owner if key backup is required by the end user or the CSP. The CA is also responsible for key recovery and may support key escrow, which allows some other entity, such as a government agency, to access the private keys issued by the CA. Key backup and key escrow are only applied to decryption keys.

In addition to the certificates which the CA issues to end users or other CAs, it also issues certificates to itself. They are called self-issued certificates. In self-issued certificates, the subject distinguished name and the issuer distinguished name are identical. Self-issued certificates are used to communicate information about the CA. For example, if the CA changes its key pair, then the CA issues a certificate with the public key of the new key pair, which is signed with the private key of the old key pair. A reason for the key pair to be changed may be that the size of the old key is no longer considered to be secure. More information about key update can be found in [2].

A special case of self-issued certificates is self-signed certificates. They certify the public key which must be used to verify the signature on the certificate itself.

They are also called root certificates and are used as trust anchors in certification paths.

Since the CA is a very security-sensitive component of the CSP, the CA is typically protected by several measures. Examples of this are key sizes that are larger than the sizes of the end user keys, protection of the private key of the CA in a hardware security module, and offline operation of the CA. Also, organizational protection mechanisms may be used, such as access control that requires several entities to collaborate and keeping the CA location secret.

7.4 Other Components

In addition to the registration authority and the certification authority, the CSP may have other components.

The CSP may operate a dedicated directory service. This service publishes certificates and delivers certificates and PSEs that contain private keys to the users. The directory service may also publish certificate revocation lists. For example, such a directory service is useful if the CA operates offline. As already discussed in Chap. 5, further components may be a *revocation service* that issues indirect certificate revocation lists or an OCSP server.

Another useful CSP component is a *time-stamping authority* (TSA). A TSA certifies that a document existed at a certain point in time by signing the document with the current date and time. They are specified in [1]. For example, TSAs are used for long-term archiving of documents as described in [6].

Finally, servers for the server-based certificate validation protocol may construct and validate certification paths on behalf of users. See Sect. 9.5 for details.

7.5 Communication Within CSPs

The efficiency and security of a PKI relies on the efficiency and security of the communication between the CSP components and between the CSP and the PKI users. In this section we describe standardized formats and protocols that support this communication.

7.5.1 *Cryptographic Protection of Messages*

Data that is communicated in a PKI requires cryptographic protection. For example, it may be encrypted or digitally signed. A common format for protecting data cryptographically is PKCS#7, which is specified in [8] and [4].

```
ContentInfo ::= SEQUENCE {
  contentType ContentType,
  content [0] EXPLICIT ANY DEFINED BY contentType OPTIONAL }
```

Listing 7.1 The ASN.1 structure of PKCS#7

The ASN.1 specification of a PKCS#7 structure is presented in Listing 7.1. PKCS#7 files are containers for *content* elements which are cryptographically protected. The `contentType` determines the form of cryptographic protection of the data. Listing 7.2 shows the six PKCS#7 content types with their respective OIDs.

The content type *data* contains plain data that is not cryptographically protected. The type *enveloped-data* consists of encrypted content of any type and the corresponding encryption keys. These encryption keys are encrypted themselves using the public keys of one or more recipients. This type can be viewed as a digital envelope which can only be opened by designated receivers. The type *signed-data* contains data of any type and one or several digital signatures on this data. To allow the transport of certificates and certificate revocation lists, this type even allows there to be no digital signature. The combined type *signed-and-enveloped data* contains signed data that are enveloped. To generate such content, first, content of type signed-data is generated, which is then enveloped. The content type *digested-data* contains data of any type together with a hash value of the data and a description of the hash function that has been used to calculate the hash value. Data of type *encrypted-data* is encrypted. However, in contrast to enveloped data there is no encryption key. Keys are assumed to be managed by other means.

The *cryptographic message syntax* (CMS) is another format for protecting data cryptographically. It is specified in [3]. CMS is based on PKCS#7. The content type signed-and-enveloped-data has been removed. Instead, in CMS the content types signed-data and enveloped-data can be nested. Also, a content type *authenticated-data* with OID "1.2.840.113549.1.9.16.1.2" is added. It contains data that are supplemented by a MAC value.

In addition to the use in CSPs, CMS is also used in many other applications such as email security [7] or for long-time archiving (see [6]).

7.5.2 Certificate Requests

In this section we describe formats for sending certificate requests to CSPs. Such formats are used in the registration process.

The certification request format PKCS#10 is widely used. It is specified in [10] and [5]. Listing 7.3 shows important parts of the ASN.1 description of PKCS#10.

A PKCS#10 file consists of a certificate request, a signature on this request, and the algorithm used to create this signature. The certificate request contains the

```
contentType: 1.2.840.113549.1.7.1

Data ::= OCTET STRING

contentType: 1.2.840.113549.1.7.3

EnvelopedData ::= SEQUENCE {
  version Version,
  recipientInfos RecipientInfos,
  encryptedContentInfo EncryptedContentInfo }

contentType: 1.2.840.113549.1.7.2

SignedData ::= SEQUENCE {
  version Version,
  digestAlgorithms DigestAlgorithmIdentifiers,
  contentInfo ContentInfo,
  certificates [0] IMPLICIT ExtendedCertificatesAndCertificates
      OPTIONAL,
  crls [1] IMPLICIT CertificateRevocationLists OPTIONAL,
  signerInfos SignerInfos }

contentType: 1.2.840.113549.1.7.4

SignedAndEnvelopedData ::= SEQUENCE {
  version Version,
  recipientInfos RecipientInfos,
  digestAlgorithms DigestAlgorithmIdentifiers,
  encryptedContentInfo EncryptedContentInfo,
  certificates [0] IMPLICIT ExtendedCertificatesAndCertificates
      OPTIONAL,
  crls [1] IMPLICIT CertificateRevocationLists OPTIONAL,
  signerInfos SignerInfos }

contentType: 1.2.840.113549.1.7.5

DigestedData ::= SEQUENCE {
  version Version,
  digestAlgorithm DigestAlgorithmIdentifier,
  contentInfo ContentInfo,
  digest Digest }

contentType: 1.2.840.113549.1.7.6

EncryptedData ::= SEQUENCE {
  version Version,
  encryptedContentInfo EncryptedContentInfo }
```

Listing 7.2 The six content types of PKCS#7

```
CertificationRequest ::= SEQUENCE {
  certificationRequestInfo CertificationRequestInfo,
  signatureAlgorithm AlgorithmIdentifier{{SignatureAlgorithms}},
  signature              BIT STRING  }

CertificationRequestInfo ::= SEQUENCE {
  version          INTEGER { v1(0) } (v1,...),
  subject          Name,
  subjectPKInfo SubjectPublicKeyInfo{{ PKInfoAlgorithms }},
  attributes     [0] Attributes{{ CRIAttributes }}  }
```

Listing 7.3 The ASN.1 structure of PKCS#10

```
CertReqMsg ::= SEQUENCE {
  certReq    CertRequest,
  popo        ProofOfPossession  OPTIONAL,
  regInfo    SEQUENCE SIZE(1..MAX) of AttributeTypeAndValue
       OPTIONAL }
```

Listing 7.4 The ASN.1 structure of a CRMF message

name of the requesting entity, the public key to be certified, appropriate parameters associated to the public key like the parameters of an elliptic curve, related attributes, and the PKCS#10 version of the request.

The attributes in the certificate request are used to communicate the client preferences to the CSP. Examples of such attributes are certificate extensions that the entity would like to include in its certificate or a pass-phrase that the entity has to provide in order to revoke its certificate. The possible attributes are described in [9, Sect. 5.4].

The signature on the request guarantees authenticity and integrity of the content. It also proves that the requesting entity possesses the private key corresponding to the certified public key if this public key is a signature verification key and the corresponding private key is used to sign the request. Obviously, this proof of possession does not work for encryption keys.

The *certificate request message format* (CRMF) [11], an alternative to PKCS#10, is very similar to this standard. The ASN.1 specification of CRMF is shown in Listing 7.4.

The `CertRequest` element allows the certificate user to send a template of a certificate to the CSP. This template can be used by the CSP to issue the certificate. In this element it is also possible to specify additional controls that are operations related to the certificate. For example, archiving of the private key by the CSP is such a control. Listing 7.5 shows the ASN.1 structure of a CertRequest.

```
CertRequest ::= SEQUENCE {
  certReqId      INTEGER,
  certTemplate   CertTemplate,
  controls       Controls OPTIONAL }

CertTemplate ::= SEQUENCE {
  version       [0] Version               OPTIONAL,
  serialNumber  [1] INTEGER               OPTIONAL,
  signingAlg    [2] AlgorithmIdentifier   OPTIONAL,
  issuer        [3] Name                  OPTIONAL,
  validity      [4] OptionalValidity      OPTIONAL,
  subject       [5] Name                  OPTIONAL,
  publicKey     [6] SubjectPublicKeyInfo  OPTIONAL,
  issuerUID     [7] UniqueIdentifier      OPTIONAL,
  subjectUID    [8] UniqueIdentifier      OPTIONAL,
  extensions    [9] Extensions            OPTIONAL }

OptionalValidity ::= SEQUENCE {
  notBefore   [0] Time OPTIONAL,
  notAfter    [1] Time OPTIONAL } --at least one must be present

Time ::= CHOICE {
  utcTime        UTCTime,
  generalTime    GeneralizedTime }
```

Listing 7.5 The ASN.1 structure of a CertRequest

7.5.3 Complex Message Formats and Protocols

In addition to the formats described so far, there are message formats and protocols that support many different PKI processes.

Such a format and protocol is the *XML key management specification* [15], which uses the XML syntax and the XML signature formats. XKMS consists of two parts, the *XML key information service specification* (X-KISS) and the *XML key registration service specification* (X-KRSS). X-KRSS supports registration, re-certification, certificate revocation, and key recovery. X-KISS allows us to locate and validate certificates.

Another protocol that allows implementing all typical PKI processes is the *certificate management protocol* (CMP), which is specified in [2]. For example, CMP supports registration, certification, revocation, management of CA keys, root CA keys and certificates, key recovery, certificate delivery and publication, and proof of possession of private keys. The specification of a CMP message can be seen in Listing 7.6.

A third complex format is *certificate management over CMS* (CMC), which is specified in [12]. The CSP clients send CMC requests. The CSP server replies with CMC responses. There are two kinds of CMC requests. A *Simple PKI Request* is simply a PKCS#10 request. A *Full PKI Request* consists of several certificate

```
PKIMessage ::=SEQUENCE {
  header            PKIHeader,
  body              PKIBody,
  protection    [0] PKIProtection OPTIONAL,
  extraCerts    [1] SEQUENCE SIZE (1..MAX) OF CMPCertificate
      OPTIONAL  }

PKIMessages ::= SEQUENCE SIZE (1..MAX) OF PKIMessage
```

Listing 7.6 The ASN.1 structure of a CMP message

Fig. 7.1 Certificates 1 and 2

Certificate 1	
Serial No.:	2593
Issuer:	CN=University CA, O=TU Darmstadt, C=DE
NotBefore:	2011-06-02
NotAfter:	2012-06-01
Subject:	CN=Bob, O=TU Darmstadt, C=DE
Public Key:	key-0x3EEA98E6
X509v3Extensions:	
Subject Alternative Name: email:bob@tu-darmstadt.de	
KeyUsage: critical digitalSignature	
Subject Key Identifier: keyId: 99:13:D5:FD:90:31:7B:56:7F:BD	
Authority Key Identifier: keyId: 0B:F8:2B:B7:B5:88:C8:03:7E:EB aci: CN=Master CA, O=TU Darmstadt, C=DE acsn: 03	

Certificate 2	
Serial No.:	2594
Issuer:	CN=University CA, O=TU Darmstadt, C=DE
NotBefore:	2011-06-02
NotAfter:	2012-06-01
Subject:	CN=Alice, O=TU Darmstadt, C=DE
Public Key:	key-0x704400D7
X509v3Extensions:	
Subject Alternative Name: email:alice@tu-darmstadt.de	
KeyUsage: critical dataEncipherment	
Subject Key Identifier: keyId: A2:F4:67:23:28:C2:C8:64:A8:45	
Authority Key Identifier: keyId: 0B:F8:2B:B7:B5:88:C8:03:7E:EB aci: CN=Master CA, O=TU Darmstadt, C=DE acsn: 03	

requests which use PKCS#10, CRMF, or some other format. These requests are sent and protected inside a CMS container. There are also two kinds of CMC responses. A *Simple PKI Response* is a CMS container which is used to send certificates to the clients. If additional information needs to be sent to the client, a *Full PKI Response*

Fig. 7.2 Certificates 3 and 4

Certificate 3	
Serial No.:	2595
Issuer:	CN=University CA, O=TU Darmstadt, C=DE
NotBefore:	2012-04-25
NotAfter:	2013-04-25
Subject:	CN=Carl, O=TU Darmstadt, C=DE
Public Key:	key-0x6D099789
X509v3Extensions:	
Subject Alternative Name: email:carl@tu-darmstadt.de	
KeyUsage: critical dataEncipherment	
Subject Key Identifier: keyId: 67:65:AE:DB:3E:2A:4C:5F:99:37	
Authority Key Identifier: keyId: 0B:F8:2B:B7:B5:88:C8:03:7E:EB aci: CN=Master CA, O=TU Darmstadt, C=DE acsn: 03	

Certificate 4	
Serial No.:	1
Issuer:	CN=Master CA, O=TU Darmstadt, C=DE
NotBefore:	2009-12-01
NotAfter:	2014-12-01
Subject:	CN=Master CA, O=TU Darmstadt, C=DE
Public Key:	key-0xB1BB55C3
X509v3Extensions:	
Basic Constraints: critical CA: TRUE pathlen: 1	
KeyUsage: critical keyCertSign, cRLSign	
Subject Key Identifier: keyId: 54:23:AA:12:AB:CD:4E:5F:3A:11	
Authority Key Identifier: keyId: 54:23:AA:12:AB:CD:4E:5F:3A:11 aci: CN=Master CA, O=TU Darmstadt, C=DE acsn: 01	

is used which can also contain other data. In addition, CMC offers mechanisms to prove the user identity and possession of private keys.

The techniques that can be used to transfer CMC messages are specified in [13]. One option is to use the POST method of HTTP(S). A second option is to use special MIME types and send the messages by email. A third option is to store the messages in a file. This is a common technique when some components are offline. By using the proper extension it is possible to specify the message type. The possible extensions are `.p10` for a Simple PKI Request, `.crq` for a Full PKI Request, `.p7c` for a Simple PKI Response, and `.prp` for a Full PKI Response. It is also possible to send the message as a binary string using TCP.

The usage of CMC is simplified when the compliance requirements specified in [14] are respected. For example, these requirements refer to the request and response types that are supported or the cryptographic algorithms that may be used.

Fig. 7.3 Certificate 5

Certificate 5	
Serial No.:	3
Issuer:	CN=Master CA, O=TU Darmstadt, C=DE
NotBefore:	2009-12-01
NotAfter:	2014-12-01
Subject:	CN=University CA, O=TU Darmstadt, C=DE
Public Key:	key-0x556486AB
X509v3Extensions:	
Basic Constraints: critical CA: TRUE pathlen: 0	
KeyUsage: critical keyCertSign, cRLSign	
Subject Key Identifier: keyId: 0B:F8:2B:B7:B5:88:C8:03:7E:EB	
Authority Key Identifier: keyId: 54:23:AA:12:AB:CD:4E:5F:3A:11 aci: CN=Master CA, O=TU Darmstadt, C=DE acsn: 01	

7.6 Exercises

7.1. Answer the following questions:

1. Which of the certificates shown in Figs. 7.1–7.3 are root certificates? Which are CA certificates? How can they be recognized? Hint: consider the extensions of the certificates.
2. How can the certificate for the public key of an issuer be found?
3. Suppose that a PKI client cannot interpret the key usage extension. How does the client treat Certificate 1?
4. Bob has generated the key pair for Certificate 1 himself. What is required by the CA to generate Bob's certificate? How can the CA convince itself that Bob knows the private key corresponding to the certified public key?
5. How can a proof of possession for Certificate 3 be performed? How does it differ from the PoP of the previous question?
6. Why is PoP important? What attacks are possible without PoP?
7. What certificates might be subject to key backup?

References

1. C. Adams, P. Cain, D. Pinkas, R. Zuccherato, Internet X.509 public key infrastructure time-stamp protocol (TSP), in *IETF Request for Comments*, 3161, Aug 2001
2. C. Adams, S. Farell, T. Kause, T. Mononen, Internet X.509 public key infrastructure certificate management protocol (CMP), in *IETF Request for Comments*, 4210, Sept 2005
3. R. Housley, Cryptographic message syntax (CMS), in *IETF Request for Comments*, 3852, July 2004

4. B. Kaliski, PKCS #7: cryptographic message syntax – Version 1.5, in *IETF Request for Comments*, 2315, Mar 1998
5. M. Nystrom, B. Kaliski, PKCS #10: certification request syntax specification – Version 1.7, in *IETF Request for Comments*, 2986, Nov 2000
6. D. Pinkas, N. Pope, J. Ross, CMS advanced electronic signatures (CAdES), in *IETF Request for Comments*, 5126, Feb 2008
7. B. Ramsdell, S. Turner, Secure/multipurpose internet mail extensions (S/MIME) Version 3.2 message specification, in *IETF Request for Comments*, 5751, Jan 2010
8. RSA Laboratories, PKCS #7 v1.5: cryptographic message syntax standard (1993), http://www.rsa.com/rsalabs/node.asp?id=2129
9. RSA Laboratories, PKCS #9 v2.0: selected object classes and attribute types (2000), http://www.rsa.com/rsalabs/node.asp?id=2131
10. RSA Laboratories, PKCS #10 v1.7: certification request syntax standard (2000), http://www.rsa.com/rsalabs/node.asp?id=2132
11. J. Schaad, Internet X.509 public key infrastructure certificate request message format (CRMF), in *IETF Request for Comments*, 4211, Sept 2005
12. J. Schaad, M. Myers, Certificate management over CMS (CMC), in *IETF Request for Comments*, 5272, June 2008
13. J. Schaad, M. Myers, Certificate management over CMS (CMC): transport protocols, in *IETF Request for Comments*, 5273, June 2008
14. J. Schaad, M. Myers, Certificate management messages over CMS (CMC): compliance requirements, in *IETF Request for Comments*, 5274, June 2008
15. World Wide Web Consortium (W3C), XML key management specification (XKMS) (2001), http://www.w3.org/TR/xkms/

Chapter 8
Certificate Policies

In the previous chapter we introduced CSPs that manage the life cycle of certificates in a hierarchical PKI. CSPs follow certain rules which are called *certificate policies*. They determine the applicability of a certificate to a particular community or class of applications with common security requirements. Therefore, certificate policies that a CSP applies while generating and managing certificates are made explicit and available to the certificate users. In addition, the CSP may describe the implementation of the policy in a *certification practice statement* (CPS). In this chapter we discuss certificate policies and certification practice statements.

8.1 Structure of Certificate Policies

RFC 3647 [2] specifies the following structure for certificate policies.

Introduction The introduction contains the name and OID of the policy. It provides definitions and terminology and defines PKI entities such as certification authorities, registration authorities, and subscribers who receive certificates from CAs. It lists applications of the certificates that comply with this policy and provides information about the administration of the policy.

Publication and Repository Responsibilities During the life cycle of a certificate, information such as the certificate policy, the certification practice statement, the certificate itself, and revocation information is published. This section of the policy contains rules on how, when, and how frequently this is done and how the published information can be accessed.

Identification and Authentication In many phases of the certificate life cycle, entities are required to identify themselves to the CSP. Examples are certificate or revocation requests. Also, the communication between those entities and the CSP must be authentic. This section specifies identification and authentication mechanisms that are being used by the CSP.

J.A. Buchmann et al., *Introduction to Public Key Infrastructures*,
DOI 10.1007/978-3-642-40657-7_8,

Certificate Life Cycle Operational Requirements This part of the policy deals with the implementation of the certificate life cycle. The policy describes the potential applicants and the application process. In particular, the policy must specify whether the applying entity may generate its own key pair and how the public key is delivered to the CSP. In addition, the policy contains requirements concerning application processing by the CSP. For example, the policy contains requirements for a registration process in which the identity of the applying entity is checked and verified. This part of the policy also specifies the requirements for certificate generation, for example, whether the certificate use is to be restricted. Also, this section contains requirements concerning notification of the certificate owners after certificate generation and acceptance of the certificates by their owners. It provides rules for dealing with expired certificates, for their renewal, and for their modification. It also contains requirements for revocation. Examples are acceptable reasons for revocation, maximum revocation latencies, and CRL issuing frequencies. Also, this section addresses processes that deal with certificate suspension, status queries, key escrow, key recovery, and users who cease to be part of the PKI.

Facility, Management, and Operational Controls For certificates to be secure, the security of the underlying cryptography and processes is not sufficient. It is also necessary to monitor and protect the facilities of the CSP and its management and to establish adequate operational controls.

Facility protection issues that are addressed by this section of the policy include physical access control, the location of the facilities and their protection, for example from water or fire, the power supply, and cleaning procedures. This part of the policy contains provisions concerning the personnel that operate the CSP. Issues are, for example, qualification, training and access to information.

In addition, this section of the policy includes provisions for logging and the evaluation and archiving of the logs. Also, CA key renewal in case of compromise or disaster is addressed, as well as possible cessation of operation of the CSP.

Technical Security Controls This part of the policy deals with secure key management, in particular the management of private keys. These mechanisms apply to the keys of the issuing CA as well as the keys of the subscribers of the CSP. This includes key generation, private key protection in personal security environments, key transport and activation, and archiving. It may also include provisions for securing the computing environment such as firewalls and secure software development for the CSP.

Certificate, CRL, and OCSP Profiles Here, the format of certificates, certificate revocation lists, and OCSP messages is specified. This information includes version, cryptographic algorithms, extensions and their criticality, policies, and policy constraints.

Compliance Audit and Other Assessment This section explains the required auditing procedures. It specifies the auditing subject, the auditing frequencies, the entities responsible for performing the audits, and provisions on how to deal with the results of audits.

Other Business and Legal Matters This policy section includes information on fees, financial responsibilities, treatment of private and confidential information, intellectual property rights, and liability.

8.1.1 Certification Practice Statement

A certification practice statement (CPS) describes how a CSP implements and enforces its chosen policy. Consequently, there may be several different certification practice statements for each certificate policy. The need for such a statement depends on the level of detail that is presented in the certificate policy itself. Certification practice statements and their relation to certificate policies are discussed in [2].

8.2 Relevant Certificate Extensions

8.2.1 Certificate Policies

The *CertificatePolicies* extension in a certificate indicates the policies that are applied by the issuer of the certificate. This extension is a sequence of OIDs and the corresponding qualifiers. The sequence may also contain the OID "2.5.29.32.0", which means `anyPolicy`.

Qualifiers provide additional information about the policy. Possible qualifiers are a pointer (URI) to the CPS document that corresponds to this policy or an explanatory text of less than 200 characters. Listing 8.1 presents the format of certificate policies extension.

In Fig. 8.1, an X.509 certificate as shown by the Windows OS can be seen. The policy of this certificate is "1.2.3.4.5.6.7.8". If a user clicks the "Issuer Statement" button on the lower right corner, he or she can see the CPS of the certificate issuer, which is typically published as a Web page.

8.2.2 Policy Mappings

The *PolicyMappings* extension is a critical extension which is set only in CA certificates. In this extension, the CA declares its own `issuerDomainPolicy`

```
CertificatePolicies ::= SEQUENCE SIZE (1..MAX) OF
   PolicyInformation

PolicyInformation ::= SEQUENCE {
  policyIdentifier   CertPolicyId,
  policyQualifiers   SEQUENCE SIZE (1..MAX) OF
       PolicyQualifierInfo OPTIONAL }

CertPolicyId ::= OBJECT IDENTIFIER

PolicyQualifierInfo ::= SEQUENCE {
  policyQualifierId  PolicyQualifierId,
  qualifier          ANY DEFINED BY policyQualifierId }
```

Listing 8.1 Certificate policies extension

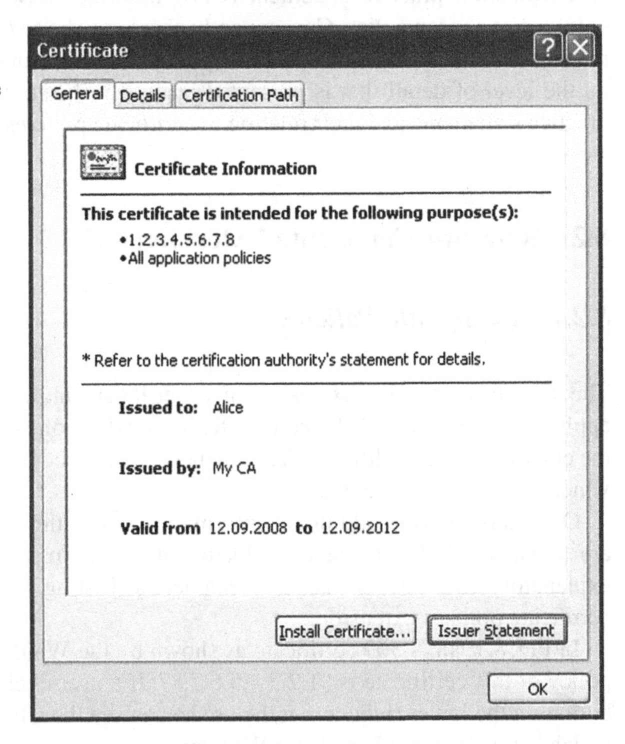

Fig. 8.1 "Issuer Statement" button on the *lower right* corner of Windows certificate viewer

to be covered by a certain `subjectDomainPolicy`. The extension contains the pair (issuerDomainPolicy, subjectDomainPolicy). Policy mapping is relevant in certification paths where policies of an issuer restrict the acceptability of policies in subsequent certificates in the path (see Sect. 9.4.1). Policy mapping is also important in cross-certification, if the cross-certifying CAs do not share the same policy.

```
PolicyMappings ::= SEQUENCE SIZE (1..MAX) OF SEQUENCE {
  issuerDomainPolicy       CertPolicyId,
  subjectDomainPolicy      CertPolicyId }
```

Listing 8.2 Policy mappings extension

When the policy mappings extension is present then the issuerDomainPolicy remains only a valid policy in the certification path if the extension contains a mapping of this policy to itself.

There are two restrictions regarding the use of this extension. It is not permitted to map anyPolicy to another policy, and vice versa. Also, any issuerDomainPolicy that is mapped to some subjectDomainPolicy should appear in the certificate policy extension. Listing 8.2 presents the format of the policy mappings extension.

8.2.3 Policy Constraints

This extension may only be set in CA certificates. It restricts the use of policies in certificates in a certification path that follow the certificate in which the extension is present.

There are two types of policy constraints, *requireExplicitPolicy* and *inhibitPolicyMapping*, that are both represented by nonnegative integers. A policy constraint of the first type is the number of certificates that may follow in the path before the presence of an explicit policy is mandatory in all further certificates in the path. The other type restricts the use of policy mapping. It contains the number of certificates that may follow in the path before policy mapping is prohibited. If a CA requires an explicit policy to be present and wants to prohibit policy mapping in all subsequent certificates, it sets both policy constraint values to 0. The path validation algorithm given by RFC 5280 in Sect. 6 of [3] (cf. Sect. 9.4.1) does not process policy constraints of self-signed certificates provided as trust anchors, but explicitly allows implementations to do so. Self-issued certificates are always processed.

8.2.4 Inhibit anyPolicy

This critical extension, which may be set only in CA certificates, is used to prohibit the use of anyPolicy in certification paths. It contains the number of certificates that may follow the CA certificate in the path before the use of anyPolicy is prohibited in further certificates. Self-issued CA certificates in the path are not counted.

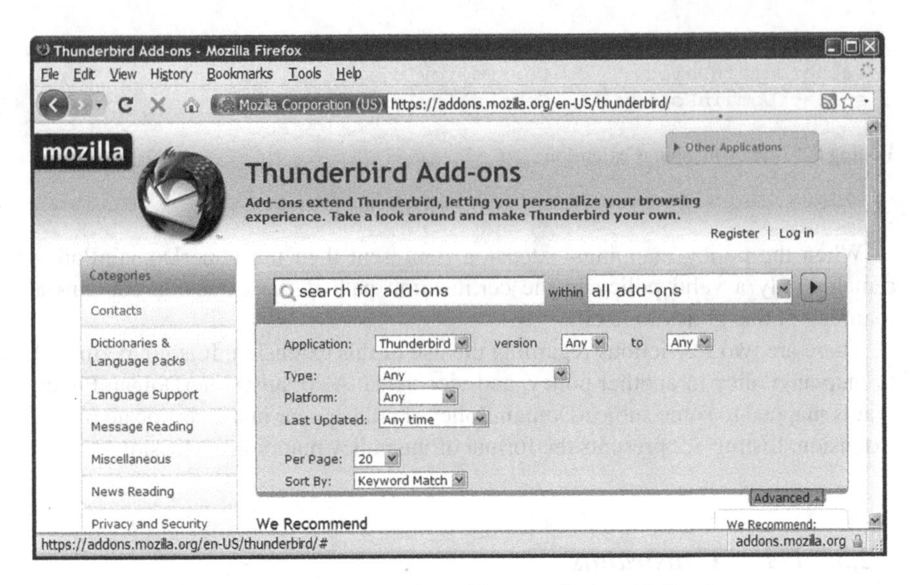

Fig. 8.2 A Web page authenticated by an EV certificate as displayed in Firefox

8.3 Extended Validation Certificates

Extended validation (EV) certificates are used to enable secure communication with Web sites. They must only be issued after extensive verification of the requesting entity's identity by the issuing CA. Extended validation certificates are specified by the Certification Authority Browser Forum [1], a voluntary consortium of certification authorities and Internet browser software and application providers that create and use SSL certificates. Issuing EV certificates requires respecting certain rules that support the security of such certificates. For example, the period between publishing CRLs must not exceed 7 days.

The use of such certificates is indicated in most browsers by displaying the corresponding URL on a green background. Figure 8.2 shows an example.

8.4 Exercises

8.1. Which of the certificates shown in Figs. 8.3 and 8.4 has a valid policy mappings extension?

Certificate A	
Serial No.:	8735
Issuer:	CN=Root CA
NotBefore:	2010-01-01
NotAfter:	2013-01-01
Subject:	CN=CA1
Public Key:	key-0xE36E2A58
X509v3Extensions:	
Basic Constraints: critical	CA: TRUE pathlen: 1
KeyUsage: critical	keyCertSign
Certificate Policies: critical	green, blue, ANY
Policy Mappings: critical	magenta ⟶ blue green ⟶ blue

Certificate B	
Serial No.:	8576845
Issuer:	CN=Root CA
NotBefore:	2010-01-01
NotAfter:	2013-01-01
Subject:	CN=Alice
Public Key:	key-0xDCFE8299
X509v3Extensions:	
Basic Constraints: critical	CA: FALSE
KeyUsage: critical	digitalSignature
Certificate Policies: critical	red, green
Policy Mappings: critical	green ⟶ blue

Fig. 8.3 Certificates A and B with policy mappings

Certificate C	
Serial No.:	8735
Issuer:	CN=Root CA
NotBefore:	2010-01-01
NotAfter:	2013-01-01
Subject:	CN=CA1
Public Key:	key-0x1FD9DB9E
X509v3Extensions:	
Basic Constraints: critical	CA: TRUE pathlen: 2
KeyUsage: critical	keyCertSign
Certificate Policies: critical	green, blue, ANY
Policy Mappings: critical	green ⟶ blue

Certificate D	
Serial No.:	907865
Issuer:	CN=Root CA
NotBefore:	2010-01-01
NotAfter:	2013-01-01
Subject:	CN=CA1
Public Key:	key-0x80B35720
X509v3Extensions:	
Basic Constraints: critical	CA: TRUE pathlen: 1
KeyUsage: critical	keyCertSign
Certificate Policies: critical	green, blue, ANY
Policy Mappings: critical	ANY ⟶ yellow

Fig. 8.4 Certificates C and D with policy mappings

References

1. CA/Browser Forum, http://www.cabforum.org/
2. S. Chokhani, W. Ford, R. Sabett, C. Merrill, S. Wu, Internet X.509 public key infrastructure certificate policy and certification practices framework, in *IETF Request for Comments*, 3647, Nov 2003
3. D. Cooper, S. Santesson, S. Farrell, S. Boeyen, R. Housley, W. Polk, Internet X.509 public key infrastructure certificate and certificate revocation list (CRL) profile, in *IETF Request for Comments*, 5280, May 2008

Chapter 9
Certification Paths: Retrieval and Validation

If entities wish to use a public key for encryption or signature verification they must retrieve this key and find out to whom it belongs. If this public key has been certified within a hierarchical PKI, the corresponding certificate must be found. Typically, such a certificate is the last element of a certification path. To verify its validity, the appropriate trust anchor must be found and the certification path must be constructed and verified. In this chapter we explain how this is done.

9.1 LDAP

Most PKIs use LDAP directories to disseminate certificates and certificate revocation lists. LDAP, the *lightweight directory access protocol* [22], allows us to access directories based on the X.500 standard. The current version of LDAP is LDAPv3. We briefly explain how LDAP works.

LDAP assumes the data to be organized in *directory information trees* (DITs) as shown in Fig. 9.1. All nodes of such a tree are called *entries*. Each entry belongs to one or more *object classes* and has one or more *attributes*. Each attribute has a name and contains data. The possible attributes that an entry can have are determined by the object classes to which the entry belongs. For example, in Fig. 9.2 the entry for user Alice belongs to the object classes top, person, and pkiUser. It has four attributes containing the corresponding data. Each entry of a DIT has a *relative distinguished name* (RDN) which is constructed from some of its attributes. Frequently, the attributes common name (cn), organizational unit (ou), and domain component (dc) are used. But any attribute can be used. Also, each entry has a distinguished name. It is a sequence that starts with the RDN of the entry and is followed by the parent's DN. The DN of an entry uniquely determines this entry within its tree. We will now explain the use of LDAP for PKI in more detail.

J.A. Buchmann et al., *Introduction to Public Key Infrastructures*,
DOI 10.1007/978-3-642-40657-7_9, © Springer-Verlag Berlin Heidelberg 2013

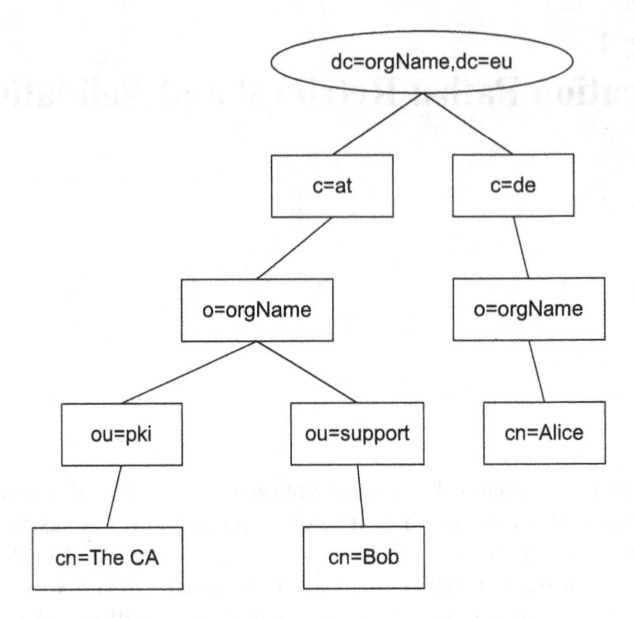

Fig. 9.1 LDAP directory information tree

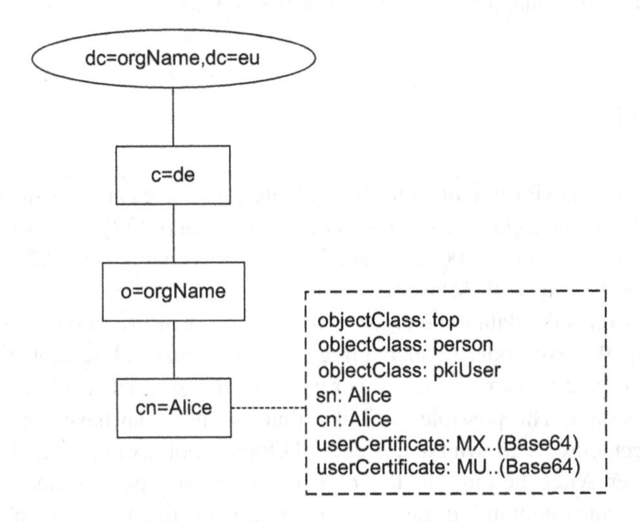

Fig. 9.2 LDAP user entry with certificates

9.1.1 Storing Certificates

User Certificates LDAP entries can be used to store certificates. For example, Fig. 9.2 shows the LDAP entry of Alice, which contains two certificates. Typically, the subjectDN of a certificate stored in an LDAP entry is equal to the DN of the entry. But this is not mandatory.

```
( 2.5.6.21 NAME 'pkiUser'
    DESC 'X.509 PKI User'
    SUP top AUXILIARY
    MAY userCertificate )
```

Listing 9.1 pkiUser object class definition

There are several object classes that permit LDAP entries to contain certificates. One of them is the object class `pkiUser` [23]. Its definition can be seen in Listing 9.1. It allows the `userCertificate` attribute to be present. This attribute may contain one or more X.509 certificates. The possibility of containing several certificates is useful when a CA issues several certificates for one user or when expired certificates are not removed from the directory. The possibility of containing no certificate permits all certificates to be removed by deleting the attribute without violating the rules of the directory.

Other object classes that allow the userCertificate attribute to be present are `strongAuthenticationUser` [23] and `inetOrgPerson` [19]. The inetOrgPerson object class is more flexible than the pkiUser object class since in addition to being able to contain user certificates it also permits providing other information about users, such as their office number or email address. In addition, this object class also permits the use of the `userPKCS12` attribute [19], which can contain a PKCS#12-encoded PSE. The object class strongAuthenticationUser is less flexible than the other two classes since it requires at least one user certificate to be present. Hence, if all certificates are removed from the userCertificate attribute, the object class must be deleted. Moreover, the strongAutheticationUser object class restricts the certificates in the corresponding entry to being used for strong authentication purposes only.

In analogy to the pkiUser object class, the `pmiUser` object class defined in [18, Clause 17] supports publication of attribute certificates by specifying the `attributeCertificate` attribute. Here, "pmi" stands for privilege management infrastructure, which manages attribute certificates just as a PKI manages public key certificates.

It is also possible for LDAP directories to store PGP certificates. Some appropriate object classes have been defined. However, they are not standardized.

CA Certificates Certificates that belong to a CA are stored in the LDAP entry of the CA which—unlike a user entry—has no other purpose. The subjectDN in such a certificate typically matches the DN of the LDAP entry. But as in the case of user certificates, this is not mandatory. As an example, Fig. 9.3 shows the entry of the CA OrgCA.

An object class that makes an LDAP entry a CA entry is `pkiCA`, defined in [23]. The specification of this object class is shown in Listing 9.2. It permits the `cACertificate` attribute, which holds one or more X.509 certificates, to be present.

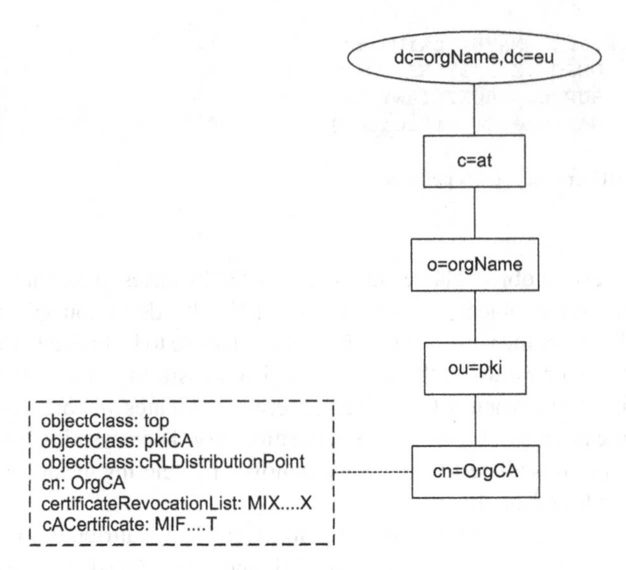

Fig. 9.3 LDAP CA entry with certificate and CRL

```
( 2.5.6.22 NAME 'pkiCA'
    DESC 'X.509 PKI Certificate Authority'
    SUP top AUXILIARY
    MAY ( cACertificate $ certificateRevocationList $
          authorityRevocationList $ crossCertificatePair $) )
```

Listing 9.2 pkiCA object class definition

```
( 2.5.6.16 NAME 'certificationAuthority'
    DESC 'X.509 certificate authority'
    SUP top AUXILIARY
    MUST ( authorityRevocationList $
           certificateRevocationList $ cACertificate )
    MAY crossCertificatePair )
```

Listing 9.3 certificationAuthority object class definition

The objectClass `certificationAuthority`, [23] which is shown in Listing 9.3, also permits CA certificates to be present. This object class is less flexible than the pkiCA object class since in addition to the cACertificate it also requires the `authorityRevocationList` attribute to be present. The latter attribute holds the authority revocation list. It is a special revocation list that exclusively lists CA certificates as revoked. Such lists are rarely used.

A cross-certificate may be stored in the entry of the CA by which it is issued and/or in the entry of the CA to which it is issued. For this purpose, the pkiCA object class permits the `crossCertificatePair` attribute to be present. It contains a pair of certificates. The issuedToThisCA certificate is the first element. The issuedByThisCA certificate is the second element. One of these elements may be absent. In previous specifications the issuedToThisCA certificate was called forward certificate and the issuedByThisCA certificate was called reverse certificate.

Attribute authority certificates are kept in the `aACertificate` attribute, which is permitted by the `pmiAA` objectClass.

9.1.2 Certificate Search

Certificates are published in LDAP directories to enable clients to access certificates that they do not already possess. Therefore, most browsers and email clients support certificate search in LDAP directories.

To search for objects in an LDAP directory, search filters are used. The format of such a filter is specified in [20]. For example, if the search filter `cn=Alice` is used, the entry shown in Fig. 9.2 is found. For entries that match the search criteria of a chosen filter, the client may specify the attributes whose values are returned by the directory. For certificate search, the userCertificate attribute can be specified. It is augmented with the option `;binary` [15] which requests that the certificates be returned in their DER-encoded form. All values of the specified attribute are returned. In our example these are two certificates. It is possible that an LDAP search via a search filter is not successful since the content of the filter is not an attribute of the corresponding LDAP entry. For example, an LDAP search that uses the filter `mail=alice@orgname.eu` to search for Alice's entry shown in Fig. 9.2 will fail because Alice's entry has no attribute containing her email address. To make the search successful, search criteria may be applied directly to the certificate instead of the entry of its owner. This is possible using *matching rules*, defined in [23], or *component matching*, described in [14]. When matching rules are applied expected certificate content is searched for in the certificates contained in the directory. For example, the `certificateExactMatch` rule [23] requires the serial number and issuerDN values provided by the searching client to match the corresponding values in a certificate stored in the directory in order for a certificate to be returned to the client. Component matching offers a more generic and flexible approach by looking for components of the ASN.1 representation of an LDAP attribute. For an example of the use of component matching for certificates, see [16]. Currently, only few clients and servers support certificate-related matching rules and component matching.

The last certificate search method that we mention is to store each certificate in a separate LDAP entry. The components of the certificate such as subjectDN or keyUsage are represented as attributes of this entry. This allows clients to locate certificates by searching for LDAP attributes. In contrast to other search techniques this method is not standardized.

A more detailed discussion of certificate search in LDAP can be found in [2].

```
( 2.5.6.19 NAME 'cRLDistributionPoint'
  DESC 'X.509 CRL distribution point'
  SUP top STRUCTURAL
  MUST cn
  MAY ( certificateRevocationList $
        authorityRevocationList $ deltaRevocationList ) )
```

Listing 9.4 cRLDistributionPoint object class definition

Table 9.1 Details of an LDAP URL

orgName.eu	Is the name of the host that operates the LDAP server
389	Is the corresponding port. It may be omitted since 389 is the default
cn=OrgCA,ou=pki,o=orgName, c=at,dc=orgName,dc=eu	Is the DN of the entry in the LDAP directory
certificateRevocationList;binary	Notifies the LDAP server to return only CRLs
base	Indicates that the search should be limited to the entry where the search starts. This makes the search very fast
objectClass=cRLDistributionPoint	Is the search filter according to [20]

9.1.3 Storing CRLs

The pkiCA object class that is used to store CA certificates also allows storing CRLs. The corresponding LDAP attribute is `certificateRevocationList`. This multi-valued attribute holds the DER-encoded CRLs. Figure 9.3 shows the entry of a CA that issues CRLs and publishes them on an LDAP server. CRLs may also be published using the `certificationAuthority` object class. However, this is not recommended due to the drawbacks described in Sect. 9.1.1. Another choice for publishing CRLs is the `cRLDistributionPoint` object class (see Listing 9.4). In addition to publishing CRLs, this object class also permits publishing delta CRLs and ARLs. This object class is a good choice for storing CRLs since it allows the cn attribute to be present, which can be used to build meaningful DNs as illustrated in Fig. 9.3. Delta CRLs are stored in the attribute `deltaRevocationList`. This object class is also useful for storing indirect CRLs since the entity related to the entry is not necessarily a CA.

If CRLs are distributed using an LDAP directory, the CRL distribution points extension (see Sect. 5.3.1) of the corresponding certificates can point to the location of the CRL in the LDAP. In the example shown in Fig. 9.3, the value of this extension is: `ldap://orgName.eu:389/cn=OrgCA,ou=pki,o=orgName,c=at, dc=orgName,dc=eu?certificateRevocationList;binary?base? objectClass=cRLDistributionPoint`. The meanings of the values in this URL (according to the format defined in [21]) are shown in Table 9.1.

9.1.4 *Security*

The LDAP protocol offers mechanisms that address various security issues. They are specified in [9]. Users can connect to the LDAP server anonymously, with their user name, or with user name and password. Anonymous connections are mostly used for LDAP searches, for example certificate searches. A more secure variant is LDAPS which uses TLS to establish a secure channel before LDAP queries start. It is also possible to combine LDAP with the *simple authentication and security layer* (SASL) architecture. The list of available SASL mechanisms can be found in [12].

LDAP also supports authorization techniques that permit users to change the content of the directory. For example, LDAP administrators may be allowed to change all data while users may only be entitled to change their own data such as password or telephone number. Moreover, certain attributes may not be visible and searchable, for example, the *userPassword* attribute that holds the password of a user.

9.2 Other Certificate Retrieval Methods

In addition to LDAP there are other methods that support the retrieval of certificates and revocation information. Examples are the domain name system [13], HTTP certificate stores [8], Web servers and FTP servers [11]. We will now discuss these possibilities in more detail.

9.2.1 *DNS*

The domain name system is a hierarchically organized naming system for computers and services on the Internet. For example, it is used for translating the symbolic name of a host (domain name) into an Internet protocol (IP) address. DNS can also be used for organizing certificates and CRLs. This is described in RFC 4398 [13]. The certificates and CRLs are stored in a *resource record* called CERT. They have four fields. The first field contains the type of the certificate or CRL, for example X.509, PGP, SPKI, and attribute certificate. The second field contains a key tag that describes the key contained in the certificate. It is a 16-bit identifier that is derived using a key tag calculation algorithm (see Appendix B of [1]). This algorithm is not guaranteed to create unique representations for keys. However, it can be used to speed up key search. The third field contains the signature algorithm used to sign the certificate or CRL as specified in Appendix A of [1], and the fourth field contains the certificate or CRL reference.

9.2.2 HTTP

RFC 4387 [8] specifies how to search for X.509 or PGP certificates and CRLs using simple HTTP requests. The RFC does not specify how those objects are stored. In fact, any repository can be used for this purpose, such as databases, LDAP directories, or even existing file systems. This search method only supports simple requests. For example, clients can search for certificate fingerprints.

9.2.3 Web Servers and FTP Servers

Another way of publishing certificates and CRLs is via HTTP or FTP servers. This is specified in [11]. The certificates and CRLs are stored in DER-encoding on the server. The respective extensions are ".cer" and ".crl". The MIME types are "application/pkix-cert" and "application/pkix-crl". Certificates and CRLs can be retrieved using the corresponding URI. Examples of such URIs are `ftp://myFtpServer.de/certificates/myCertificate.cer` and `http://myHttpServer.com/crls/myCrl.crl`. This method has several advantages. Most organizations use FTP or HTTP servers. The URI is static. This makes storing, retrieving and disseminating location information very easy.

9.2.4 WebDAV

In [3] the authors propose a method that uses the Web-based distributed authoring and versioning standard (WebDAV) [6] for publishing certificates and revocation lists. In WebDAV-based publication, every certificate and CRL is represented as a Web page. In addition, the owner of one or more certificates is a *collection* (comparable to a file system directory) and the corresponding certificates are members of this collection. The certificates and CRLs can be accessed using HTTP requests. The respective extensions are ".p7c" for certificates, ".crl" for CRLs, and ".ace" for attribute certificates. The information on where to locate the certificate is contained in the *Authority Information Access* extension (see Sect. 9.6.1). Revoked certificates can be removed from the servers. The non-existence of a certificate is an easy way of informing users about the revocation. Alternatively, a CRL with a single entry can be placed on the server.

9.3 Certification Path Building

In RFC 4158 [4], certification path building is defined as "the process used to assemble the certification path between the trust anchor and the target certificate". Certification path building is a complex process. For example, it requires clients

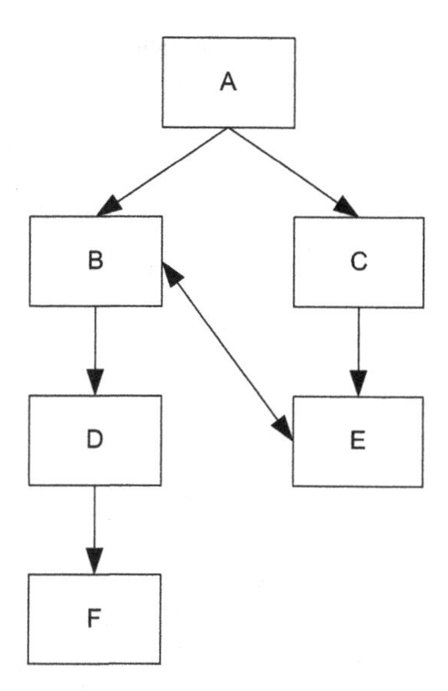

Fig. 9.4 An example of a PKI hierarchy

that construct such paths to understand many different certificate extensions and to support different protocols for accessing PKI repositories.

We explain the algorithm for certificate path building from RFC 4158. Constructing a certification path requires two inputs, the trust anchor that the path starts from and the target certificate that the path ends with. The trust anchor can be given in the form of a certificate or explicitly by its DN, a public key, an algorithm, and corresponding parameters.

We illustrate the process of certificate path building in an example. Figure 9.4 shows a PKI as a graph. It has six participants which are the nodes. Entity A is the trust anchor. An edge from X to Y means that X has issued a certificate to Y. This certificate is denoted by C_Y^X.

In this PKI there are two certification paths from A to F. They can be constructed from the trust anchor as shown in Fig. 9.5. The tree in this figure is called the `issuedBy` direction tree. They can also be constructed from the target certificate, as in Fig. 9.6, which shows the `issuedTo` direction tree. RFC 4158 calls the issuedBy direction *forward* while the issuedTo direction is referred to as *reverse*.

As can be seen in Figs. 9.5 and 9.6, there may be certification paths of different lengths. Which path is found depends on the certificates that are being used in the construction. When the client starts with certificate C_C^A, the path is longer than when it starts with certificate C_B^A.

The direction in which the certification path is constructed mainly depends on what information is available. If all cross-certificates and CA certificates are published in the directory that holds the target, then using both directions is possible. If only the issuedToThisCA certificates are published, then the issuedTo direction is used, while the opposite direction can be chosen when the issuedByThisCA certificates are present. The issuedTo direction can also be selected if all certificates contain information on how to locate the corresponding issuer certificates.

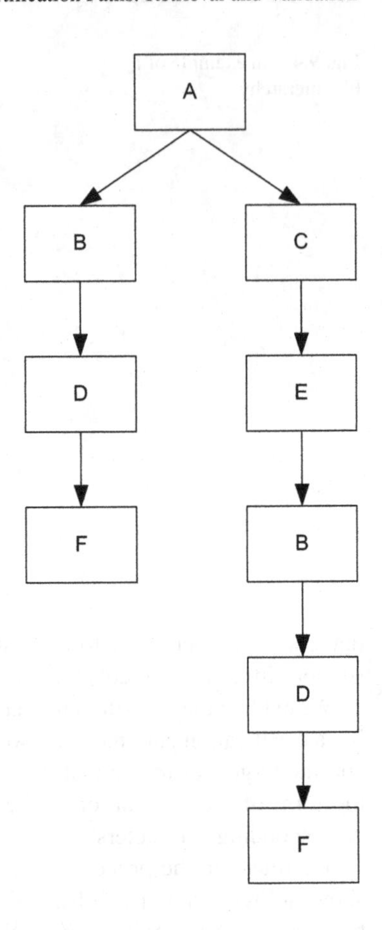

Fig. 9.5 issuedBy direction tree

If there are different certification paths, then certification path building must be optimized. For example, consider the PKI from Fig. 9.7. The shortest path from the trust anchor B that validates C_G^E is C_A^B, C_C^A, C_E^C. The path C_A^B, C_C^A, C_D^C, C_B^D, C_A^B, C_C^A, C_E^C is invalid because it contains a certificate (C_A^B) twice. However, the path C_A^B, C_C^A, C_D^C, C_F^D, C_C^F, C_E^C is valid but longer than the shortest path. This path contains two certificates for entity C. Avoiding this is a simple optimization. Methods for optimizing certification path building can be found in Sect. 3 of [4]. Certification path building can be performed simultaneously with certification path validation, which is discussed in the following section.

9.4 Certification Path Validation

In this section we explain the algorithm for certification path validation specified in RFC 5280 (Sect. 6 of [5]).

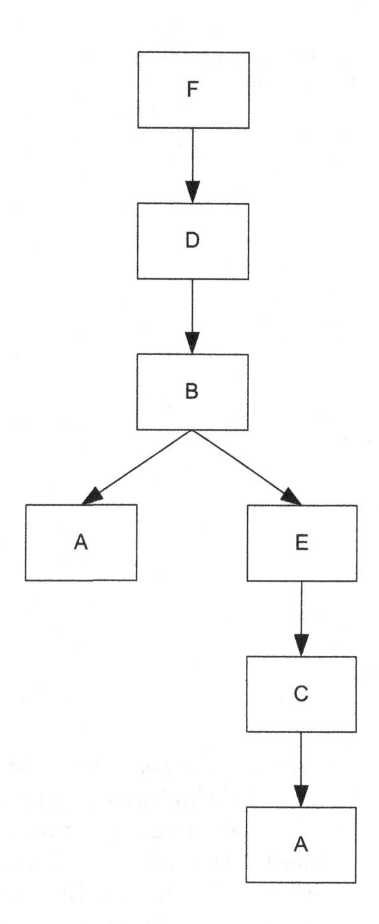

Fig. 9.6 issuedTo direction tree

9.4.1 Validation Algorithm

RFC 5280 proposes an algorithm that validates a given certification path in Sect. 6 of [5]. If the algorithm fails to validate the path, the reason for this failure is also given.

This algorithm is rather complicated and not frequently used in practice. Therefore, we only sketch the algorithm. The details can be found in Appendix A.

The goal of the algorithm is to find out whether the path is valid in the shell model and whether the policies that were used when the certificates were constructed comply with the preferences of the user. Checking the latter is responsible for the algorithm being so complex. Obviously, it would be easier if all the certificates in the path were constructed under the same policy and if the user accepts this policy. Apparently, the inventors of the algorithm found this too restrictive. Thus, the algorithm allows policies that are equivalent. The equivalence of policies is determined by the certificates in the path, which may map policies to other equivalent policies. To deal with the possibility of policy mapping, the algorithm

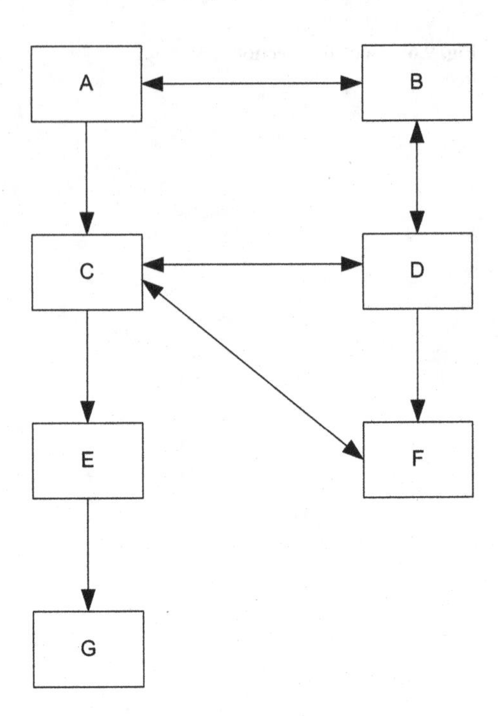

Fig. 9.7 An example of a PKI hierarchy

constructs a *valid policy tree*. The existence of a valid policy tree at the time when the algorithm terminates indicates that the certification path is valid.

The algorithm receives several inputs. One input is the certification path to be validated and the time at which it is to be validated. The last element of the path is the certificate to be validated. The trust anchor is not included in this path. It is provided separately to the algorithm. As second input, the algorithm receives information about the preferences of the user who validates the path regarding policies. For example, a set of identifiers of certificate policies is supplied that are accepted by the user. Also, information is provided whether or not the user accepts policy mapping.

The algorithm proceeds in four steps. The workflow of the algorithm is depicted in Fig. 9.8. The first step is the *initialization*, which is performed exactly once. In this step, a number of variables are initialized that are modified during the algorithm. For example, parameters for certificate verification and the valid policy tree are initialized. After the initialization the algorithm validates the certificates iteratively in the *basic certificate processing*. This step is executed as many times as there are certificates in the path unless the algorithm terminates before all certificates are processed. In this step, signature verification is performed, it is checked whether the validity period of the certificate contains the validation time, and the policy tree is updated. Each of these processes can lead to the termination of the algorithm, in which case the certification path is invalid. As long as there are more certificates to be processed, basic certificate processing is followed by *preparation for the next*

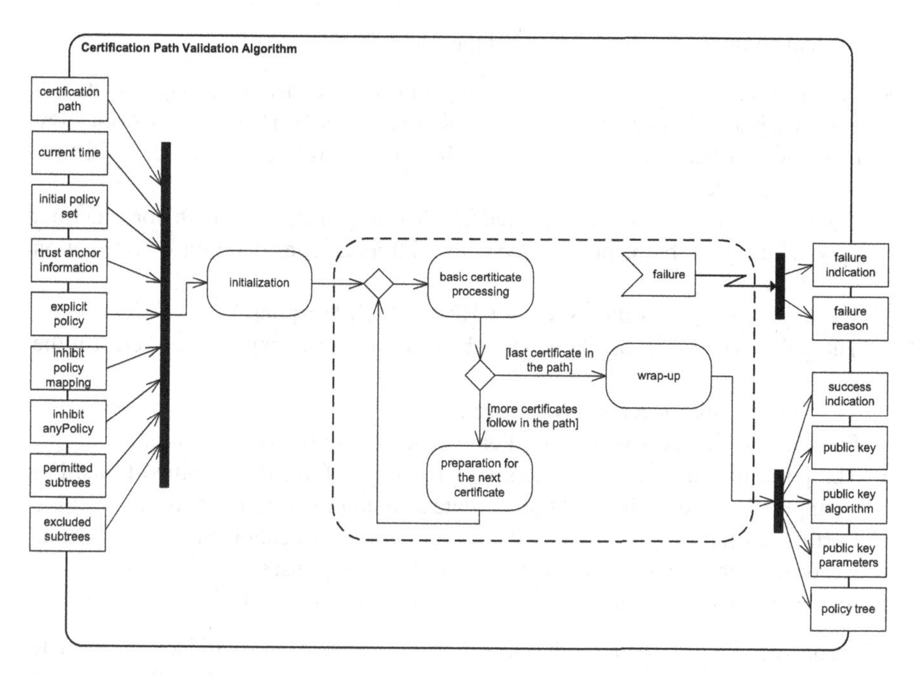

Fig. 9.8 UML activity diagram of RFC 5280 path validation algorithm

certificate. For example, in this step the parameters for verifying the validity of the next certificate are extracted and policy mapping is processed. The last step is *wrap-up*. It is performed once and gathers the output information. If the algorithm is successful, it outputs a success indicator, a valid policy tree, and the public key that has been verified to belong to the subject in the last certificate in the path.

9.5 Server-Based Certificate Validation Protocol (SCVP)

Some clients outsource certification path construction and validation because these tasks are too complex for them or they do not have access to necessary resources such as certificate archives. This is why the server-based certificate validation protocol (SCVP) has been specified in [7].

This protocol is based on a client-server architecture. The clients make a request (`CVRequest`) and get a response (`CVResponse`) from the server. For authenticity purposes, both the request and the response may be signed or authenticated by a MAC. Correspondingly, the SCVP messages are placed in a `SignedData` or an `AuthenticatedData` container, which, in turn, is placed in a `ContentInfo` message (see [10]). The messages exchanged between client and server are specified in ASN.1 and are DER-encoded.

Typical content of a client request is the following.

- The certificate for which a certification path is to be constructed and validated. It may be a public key certificate or an attribute certificate. There may be even more than one certificate of the same type. Alternatively, references of the certificates may be included.
- The tasks that the server is expected to perform: certification path construction, or validation, or both, potentially with additional consideration of revocation information.
- The expected return values such as CRLs or OCSP responses.
- The policy (see [17]) according to which the server is expected to perform the validation.
- The trust anchors accepted by the user.
- Previous SCVP responses if the client wishes an alternative response.
- The point in time for which the server should perform the specified tasks. For example, this allows the client to validate a signature at signature time.
- Certificates that may be used by the server in the certification path construction.
- Revocation information such as CRLs or OCSP responses.
- Specification of how old acceptable cached responses may be.

Upon receiving a request, the server tries to create and validate appropriate certification paths, for example, by connecting to repositories, OCSP servers, or other SCVP servers. Then it returns a response to the client.

9.6 Relevant Certificate Extensions

There are several certificate extensions that are relevant for certification path construction and validation. Two of them are described in this section.

9.6.1 Authority Information Access

The private extension *authority information access* (see [5]) of a certificate C can point to certificates that precede C in a certification path. This supports certification path building. The extension can also point to an OCSP server that provides revocation information regarding C. The extension has an `accessMethod` field that describes its content and an `accessLocation` field.

The possible access methods are `caIssuers` and `ocsp`. They are represented by the corresponding OIDs. caIssuers indicates that the extension points to certificates issued to the issuer of the certificate. The corresponding accessLocation field points to a location where these certificates can be found. The purpose of this information is to aid certificate users in the certification path construction. The ocsp access method indicates that the extension points to an OCSP server that can

Certificate A	
Serial No.:	1
Issuer:	CN=Root CA
NotBefore:	2009-01-01
NotAfter:	2014-12-31
Subject:	CN=Root CA
Public Key:	key-0x77004433
X509v3Extensions:	
Basic Constraints: critical	CA: TRUE pathlen: 2
KeyUsage: critical	keyCertSign
Certificate Policies: critical	ANY

Certificate B	
Serial No.:	2
Issuer:	CN=Root CA
NotBefore:	2009-01-02
NotAfter:	2012-12-31
Subject:	CN=CA1
Public Key:	key-0xBBDD5588
X509v3Extensions:	
Basic Constraints: critical	CA: TRUE pathlen: 1
KeyUsage: critical	keyCertSign
Certificate Policies: critical	green, blue, ANY

Certificate C	
Serial No.:	3
Issuer:	CN=CA1
NotBefore:	2010-01-01
NotAfter:	2012-12-31
Subject:	CN=CA2
Public Key:	key-0x12340987
X509v3Extensions:	
Basic Constraints: critical	CA: TRUE pathlen: 0
KeyUsage: critical	keyCertSign
Certificate Policies: critical	green, yellow

Certificate D	
Serial No.:	25
Issuer:	CN=CA2
NotBefore:	2012-01-01
NotAfter:	2012-12-31
Subject:	CN=Bob
Public Key:	key-0x80907060
X509v3Extensions:	
KeyUsage: critical	dataEncipherment
Certificate Policies: critical	yellow

Fig. 9.9 Certification hierarchy to be validated

provide status information about the certificate. The corresponding accessLocation field points to this OCSP server.

The extension may contain several access method and access location fields.

9.6.2 *Subject Information Access*

The purpose of this private extension is to provide references to locations of data and services associated to the subject of the certificate.

It has the same syntax as the authority information access extension containing access method and access location pairs. Two access methods are defined. One is `caRepository`, which is used only in CA certificates. It provides the location of a directory where the CA that is the subject of the certificate publishes certificates that it issues. The second method is `timeStamping`, which is used only in end

Table 9.2 Input variables of the algorithm

cp	
date	
uips	
public key	
ipolmap_inh	
iexpol	
iapol_inh	
ipersub	
iexsub	

Table 9.3 Variables of the algorithm and their values during execution

Variables	Iteration			
	$i = 1$	$i = 2$	$i = 3$	(Wrap-up)
v_p_t				
ex_pol				
in_ap				
pol_map				
w_pk				
w_iss				
m_path				

entity certificates that belong to time-stamping servers and provides the location of a time-stamping server.

9.7 Exercises

9.1. Consider the certificates given in Fig. 9.9. Appendix A is required for solving the exercise.

Alice verifies the authenticity of Bob's public key on 2012-03-23 using the path validation algorithm. Her trust anchor is the "Root CA".

1. Create a tree that depicts the certification hierarchy for these certificates.
2. What is the length n of the certification path cp?
3. Alice allows all policies, even the anyPolicy, but she wants the certificates to satisfy at least one explicit policy. She also allows policy mapping and does not put any constraints on the names contained in the certificates. Enter the values of the input variables in Table 9.2. See Appendix A for the meaning of the variable names.
4. Enter the states of the algorithm at the beginning of each iteration (Table 9.3).
5. Draw the valid policy tree *v_p_t*.
6. What is the output of the algorithm?

References

1. R. Arends, R. Austein, M. Larson, D. Massey, S. Rose, Resource records for the DNS security extensions, in *IETF Request for Comments*, 4034, Mar 2005
2. D. Chadwick, Deficiencies in LDAP when used to support PKI. Commun. ACM **46**(3), 99–104 (2003)
3. D.W. Chadwick, S. Anthony, Using WebDAV for improved certificate revocation and publication, in *Proceedings of Public Key Infrastructure: 4th European PKI Workshop: Theory and Practice, EuroPKI 2007*, June 2007, Palma de Mallorca. Volume 4582 of Lecture Notes in Computer Science, pp. 265–279
4. M. Cooper, Y. Dzambasow, P. Hesse, S. Joseph, R. Nicholas, Internet X.509 public key infrastructure: certification path building, in *IETF Request for Comments*, 4158, Sept 2005
5. D. Cooper, S. Santesson, S. Farrell, S. Boeyen, R. Housley, W. Polk, Internet X.509 public key infrastructure certificate and certificate revocation list (CRL) profile, in *IETF Request for Comments*, 5280, May 2008
6. L. Dusseault, HTTP extensions for web distributed authoring and versioning (WebDAV), in *IETF Request for Comments*, 4918, June 2007
7. T. Freeman, R. Housley, A. Malpani, D. Cooper, W. Polk, Server-based certificate validation protocol (SCVP), in *IETF Request for Comments*, 5055, Dec 2007
8. P. Gutmann, Internet X.509 public key infrastructure operational protocols: certificate store access via HTTP, in *IETF Request for Comments*, 4387, Feb 2006
9. R. Harrison, Lightweight directory access protocol (LDAP): authentication methods and security mechanisms, in *IETF Request for Comments*, 4513, June 2006
10. R. Housley, Cryptographic message syntax (CMS), in *IETF Request for Comments*, 5652, Sept 2009
11. R. Housley, P. Hoffman, Internet X.509 public key infrastructure operational protocols: FTP and HTTP, in *IETF Request for Comments*, 2585, May 1999
12. Internet Assigned Numbers Authority IANA, Simple authentication and security layer (SASL) mechanisms, http://www.iana.org/assignments/sasl-mechanisms/sasl-mechanisms.xml
13. S. Josefsson, Storing certificates in the domain name system (DNS), in *IETF Request for Comments*, 4398, Mar 2006
14. S. Legg, Lightweight directory access protocol (LDAP) and X.500 component matching rules, in *IETF Request for Comments*, 3687, Feb 2004
15. S. Legg, Lightweight directory access protocol (LDAP): the binary encoding option, in *IETF Request for Comments*, 4522, June 2006
16. S.S. Lim, J.H. Choi, K.D. Zeilenga, Design and implementation of LDAP component matching for flexible and secure certificate access in PKI, in *Online Proceedings of the 4th Annual PKI R&D Workshop*, Gaithersburg, Apr 2005. http://middleware.internet2.edu/pki05/proceedings/
17. D. Pinkas, R. Housley, Delegated path validation and delegated path discovery protocol requirements, in *IETF Request for Comments*, 3379, Sept 2002
18. Recommendation X.509 ITU-T, Information technology – open systems interconnection – the directory: public-key and attribute certificate frameworks, Aug 2005
19. M. Smith, Definition of the inetOrgPerson LDAP object class, in *IETF Request for Comments*, 2798, Apr 2000
20. M. Smith, T. Howes, Lightweight directory access protocol (LDAP): string representation of search filters, in *IETF Request for Comments*, 4515, June 2006
21. M. Smith, T. Howes, Lightweight directory access protocol (LDAP): uniform resource locator, in *IETF Request for Comments*, 4516, June 2006
22. K. Zeilenga, Lightweight directory access protocol (LDAP): technical specification road map, in *IETF Request for Comments*, 4510, June 2006
23. K. Zeilenga, Lightweight directory access protocol (LDAP) schema definitions for X.509 certificates, in *IETF Request for Comments*, 4523, June 2006

Chapter 10
PKI in Practice

In this chapter we present applications that use public key cryptography and PKIs. In these applications both PGP- and X.509-based infrastructures can be used. While X.509 is mostly used in commercial applications, PGP is popular in the private sector.

10.1 Internet

Important PKI-based Internet protocols are SSL and TLS. These protocols support confidential and authenticated channels between clients and servers. The current version of TLS is specified in [5], which replaces [4]. TLS is the successor of SSL 2.0 and 3.0 [11], specified by Netscape.

An important example of the application of TLS is Web-based e-commerce. A typical scenario is the following. A Web shop has an X.509 certificate and the corresponding private key which TLS uses to establish a secure channel between customers and the Web shop. Typically, this channel has two properties. The Web shop is authenticated to the customer and the communication between the Web shop and the customer is confidential. Thus, sensitive data such as name, address, and credit card number of the customer are protected from unauthorized access. TLS also supports client authentication, which also requires the clients to have a certificate. However, this feature is usually not used in e-commerce. A similar application of TLS is the protection of home banking applications. In such applications it is essential that the bank customer authenticates himself or herself to the bank. For this purpose, he or she possesses a private key and a certificate which are usually stored on a smart card provided by the bank.

Another application of TLS is HTTPS, a combination of TLS with the HTTP protocol. For example, HTTPS is used to authenticate Web pages. Figure 10.1 shows the TLS secured Web page of Springer. The fact that a Web page is HTTPS-secured is indicated by Web browsers. Firefox displays an icon in front of the URL. Clicking

J.A. Buchmann et al., *Introduction to Public Key Infrastructures*,
DOI 10.1007/978-3-642-40657-7_10, © Springer-Verlag Berlin Heidelberg 2013

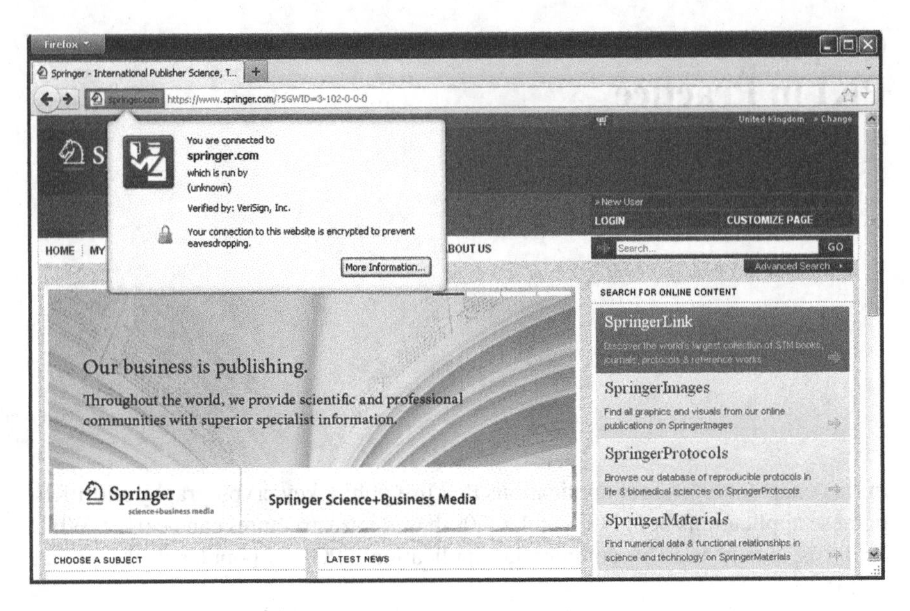

Fig. 10.1 TLS-secured Springer Web page

on the icon causes Firefox to display more information about the secure connection, as shown in Fig. 10.1. Further clicking shows even more information, such as the cryptographic algorithms used (see Fig. 10.2). Firefox and other browsers also let users examine server certificates (Fig. 10.3) and certification paths (Fig. 10.4). Such a path starts from a certificate which is directly trusted by the browser. If such a certificate does not exist, the browser displays a warning about a possibly insecure TLS connection.

TLS is also used in combination with other protocols used on the Internet, such as LDAP [13], POP3 [17], IMAP [17], SMTP [14], and FTP [10]. To address attacks against TLS, the protocol has been extended (see [21]).

Certificates that are used for TLS connections have specific content. Server certificates contain the `serverAuth` value (OID "1.3.6.1.5.5.7.3.1") in their extended key usage extension. Client certificates contain the `clientAuth` value (OID "1.3.6.1.5.5.7.3.2") in the same extension.

10.2 Email

PKI can also provide end-to-end email security. End-to-end security refers to confidentiality, authenticity, and non-repudiation. We explain two standards that provide such security.

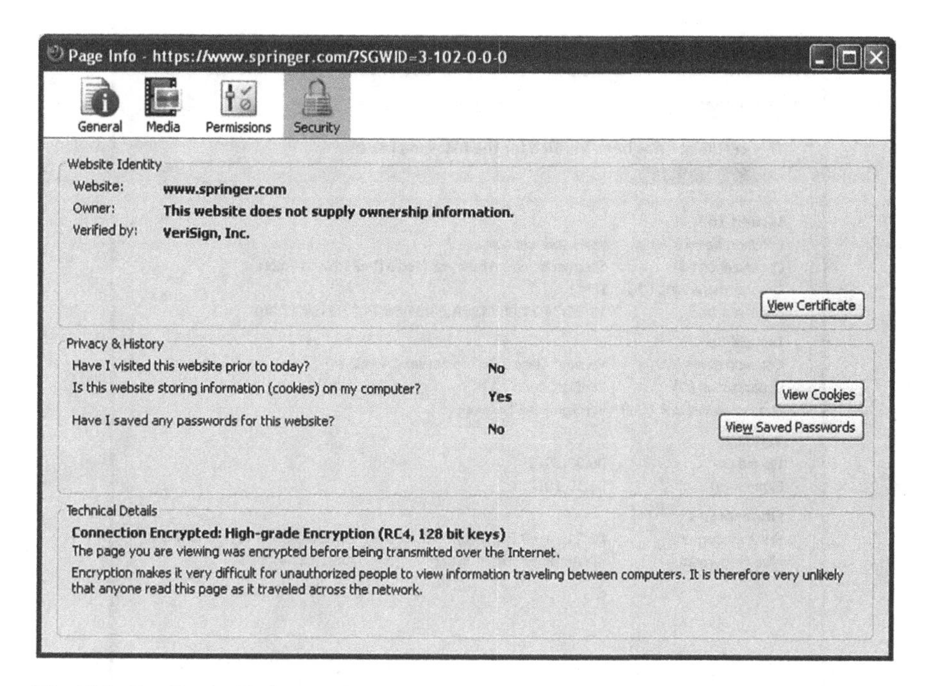

Fig. 10.2 Details of a TLS-secured Web site

10.2.1 S/MIME

One of the most important email security standards is *secure/multipurpose Internet mail extensions* (S/MIME), defined in [20]. The use of X.509 certificates in connection with S/MIME is specified in [19]. This document requires the certificate processing requirements described in [2] to be implemented by the email client and introduces some additional requirements. For example, if the certificate contains the extended key usage extension, it must contain the `emailProtection` ("1.3.6.1.5.5.7.3.4") or `anyExtendedKeyUsage` ("2.5.29.37.0") OID.

We explain how S/MIME works when the two parties Alice and Bob exchange emails using the Thunderbird email client. Both parties have a certificate issued by the certification authority CDC-CA. Figures 10.5–10.7 show the certificate categories stored by the Thunderbird certificate manager. As shown in Fig. 10.5, Alice installed her own X.509 certificate together with her private key in the category *Your Certificates*. As seen in Fig. 10.6, she installed Bob's X.509 certificate in the category *People*. Figure 10.7 also shows the category *Authorities*, where the CDC-CA root certificate is installed.

As seen in Figs. 10.8 and 10.9, working with S/MIME emails only slightly differs from working with unsecured emails once all necessary certificates are properly installed. Figure 10.8 shows Alice writing an encrypted and signed message to Bob

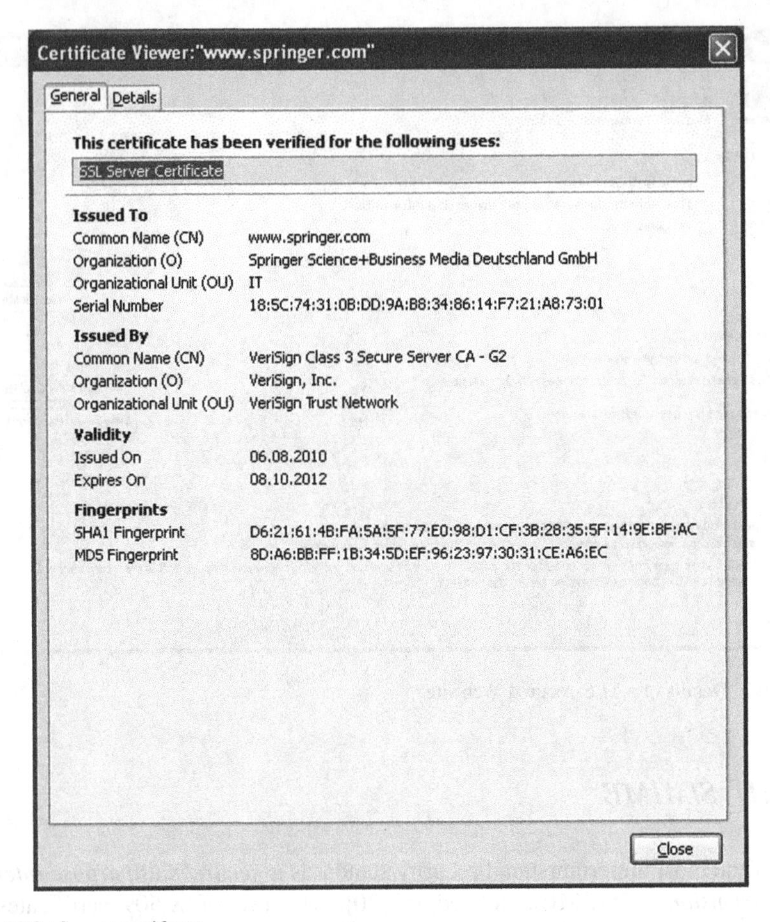

Fig. 10.3 Server certificate

using S/MIME. The only difference with writing an unsecured message is that the S/MIME menu is opened and the options *Encrypt This Message* and *Digitally Sign This Message* are selected. The current security settings are shown in the lower right corner of the email editor. Encryption is symbolized by a padlock symbol and email signature is indicated by a sealed envelope. Each of these settings can be set as a default by the user. When using S/MIME it is important to know that the email header is not encrypted even if email encryption is used. Thus, a potential eavesdropper can still see who is communicating with whom and can read the subject line.

Figure 10.9 shows Bob reading Alice's message. Encrypted emails are automatically displayed in clear but remain encrypted in the mailbox. Signed emails are automatically verified. The security settings of the email are indicated in the upper right corner of the email window. As before, an encrypted email is symbolized by

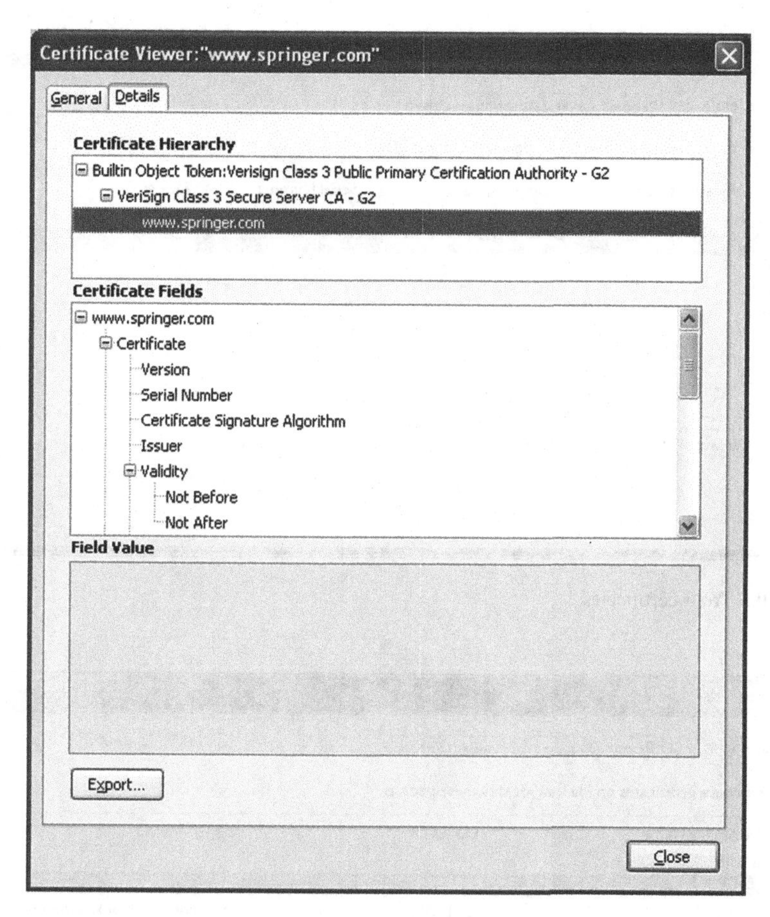

Fig. 10.4 Certification path to the server certificate (the root certificate is trusted by the browser)

a padlock and a signed email is indicated by a sealed envelope. If decryption or verification fails, a warning is displayed.

10.2.2 PGP

Another important email security standard is PGP. It is standardized as OpenPGP in [1]. The corresponding open-source implementation is GnuPG [12]. PGP email security comes in different flavors, such as PGP/INLINE as defined in [1] and PGP/MIME as defined in [6]. While PGP/INLINE includes encrypted messages and signatures in the standard email body, PGP/MIME uses a specialized email format. As this is transparent to the regular user, we do not go into details here.

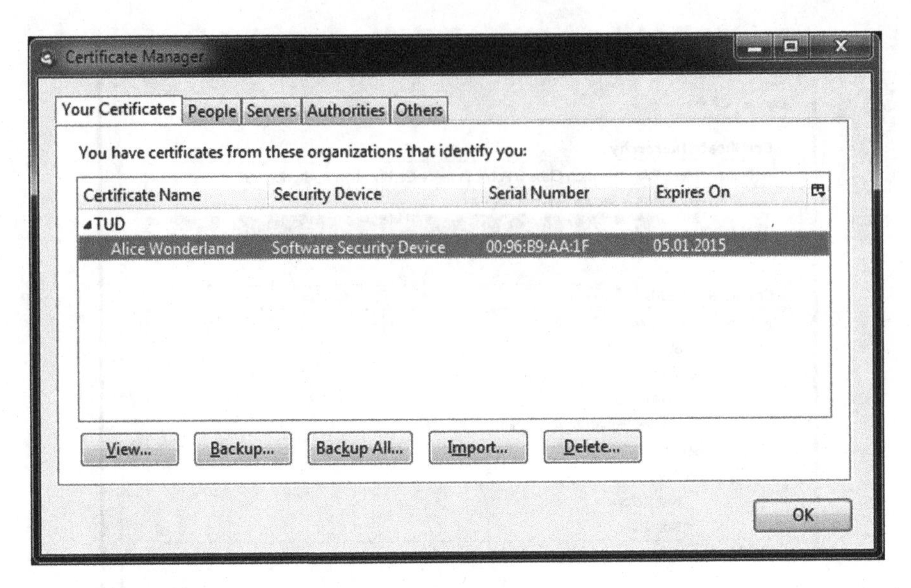

Fig. 10.5 Your certificates

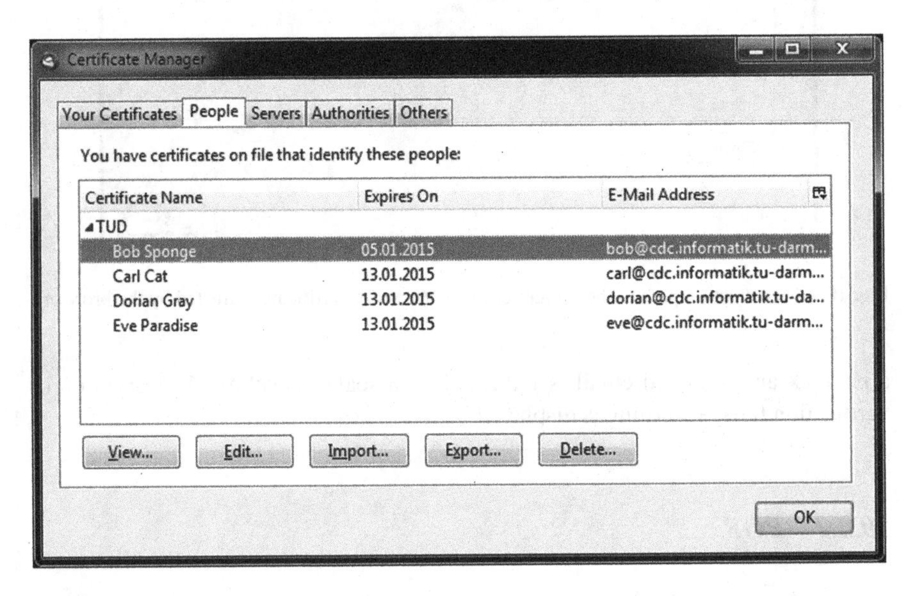

Fig. 10.6 People

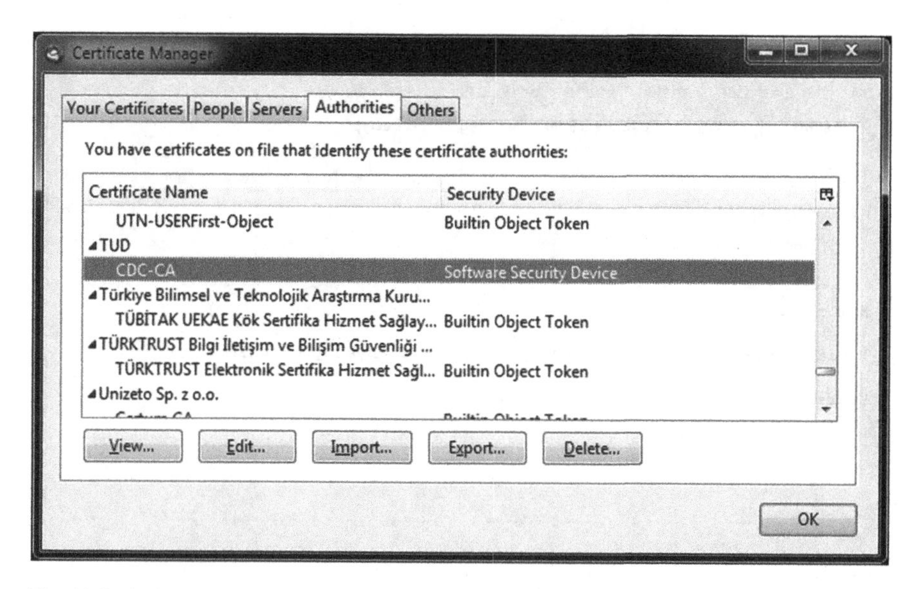

Fig. 10.7 Authorities

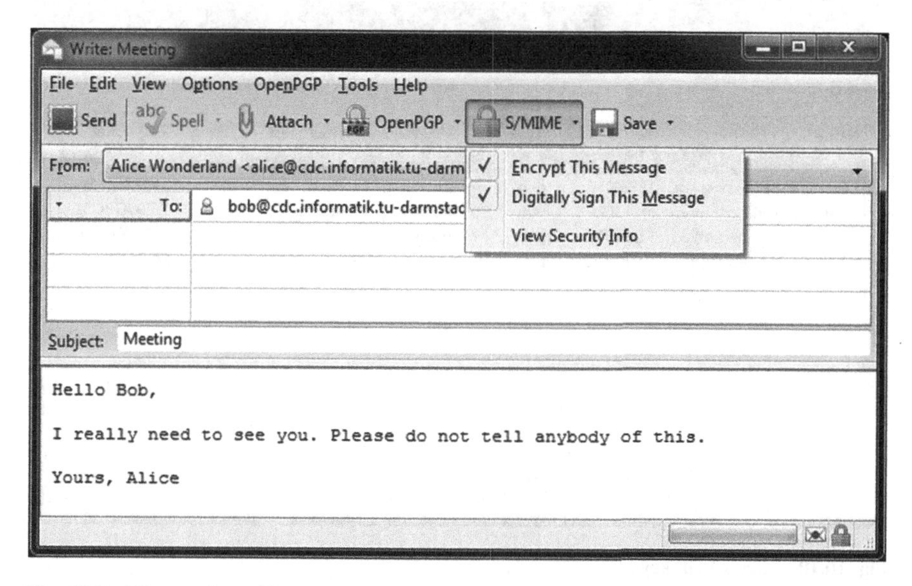

Fig. 10.8 Alice sends an S/MIME secured email

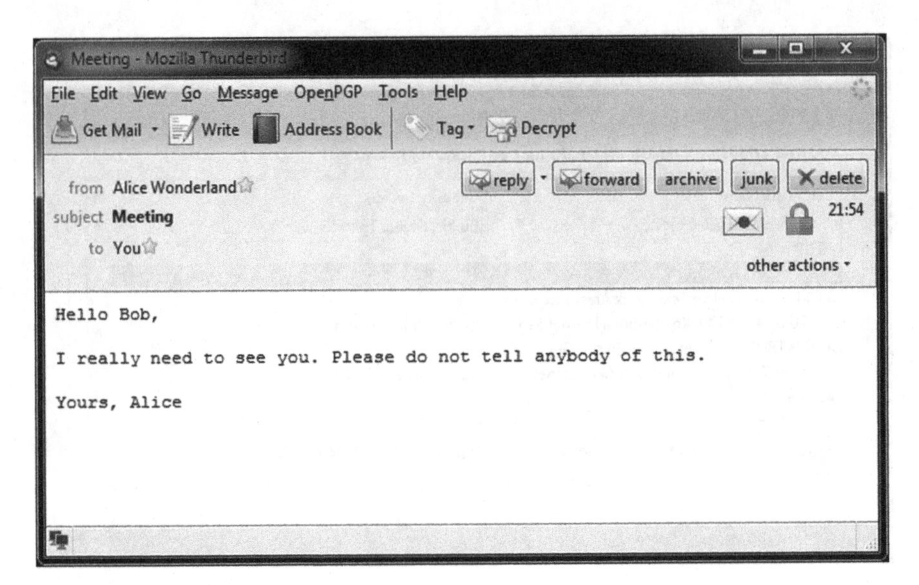

Fig. 10.9 Bob receives an S/MIME secured email

OpenPGP Key Management

File Edit View Keyserver Generate

Search for: Clear Display All Keys by Default

Name	Key ID	Type	Key Validity	Owner Trust	Expiry	Fingerprint
Alice Wonderland (no p...	F8FF1E4E	pub/sec	ultimate	ultimate		29D6 0D5C ...
▷ Bob Sponge (no passwor...	1BE3905B	pub	trusted	trusted		BA3F 8EEF 1A...
▷ Carl Cat <carl@cdc.infor...	EF815E91	pub	trusted	marginal		05C0 A7BC 51...
▷ Dorian Gray <dorian@cd...	FB28E993	pub	trusted	-		38DE 70F5 60...
▷ Eve Paradise <eve@cdc.i...	DC236D63	pub	-	-		679B C311 8B...

Fig. 10.10 Alice's PGP keys

As an example, again consider email communication between Alice and Bob using the email client Thunderbird together with EnigMail [7]. Alice and Bob both have their own and the other party's PGP certificates. Figures 10.10 and 10.11 show Alice's OpenPGP Key Management interface. As seen in Fig. 10.10, Alice has

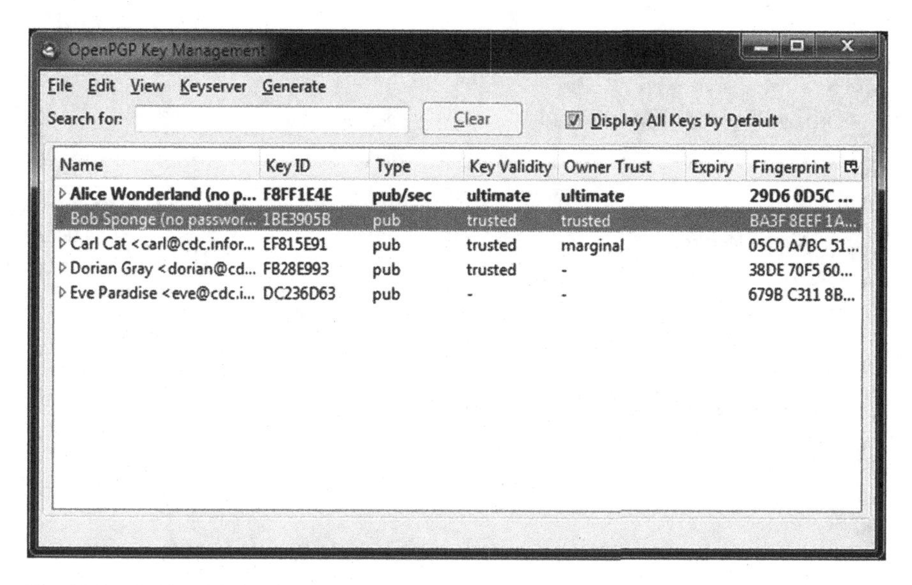

Fig. 10.11 Bob's PGP certificate

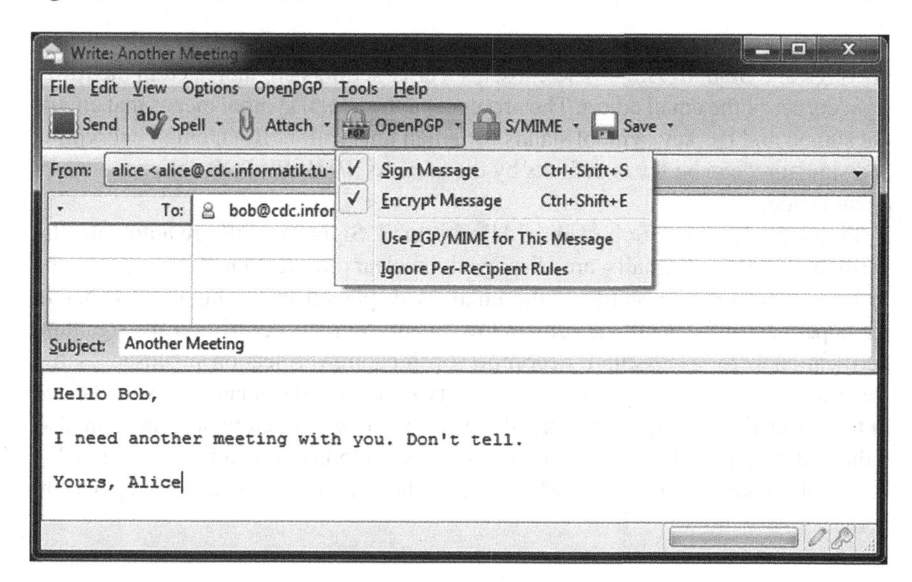

Fig. 10.12 Alice sends a PGP secured email

installed her own PGP certificate along with her private key. As seen in Fig. 10.11, Alice also installed Bob's PGP certificate.

PGP secured emails are as easy to use as unsecured emails as soon as the appropriate certificates are installed. Figure 10.12 shows Alice writing a PGP signed and encrypted email to Bob. She opens the OpenPGP menu and checks *Encrypt*

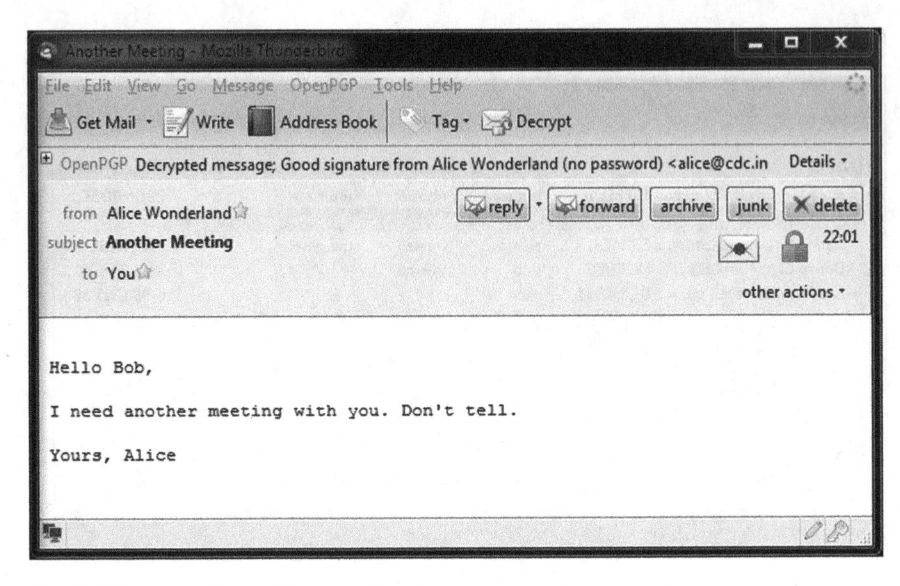

Fig. 10.13 Bob receives a PGP secured email

Message and *Sign Message*. The chosen security settings are indicated in the lower right corner of the email editor. The presence of the pencil symbol means that emails are signed and the key symbol stands for email encryption. Both, email encryption and signature, can be set as defaults by the user. As S/MIME, PGP does not encrypt email headers.

Figure 10.13 shows Bob reading Alice's email. Signed emails are automatically verified. Encrypted emails are displayed in clear but remain encrypted in the mailbox. The security status of the email is displayed in the highlighted bar in the upper part of the email viewer window. In our example, decryption and signature verification were successful. If decryption or signature verification are unsuccessful, a warning is displayed in the bar. The security settings of the email are also indicated in the lower right corner of the email viewer window. As before, a signed email is indicated by a pencil, an encrypted message is symbolized by a key. Additionally, the symbols known from S/MIME indicate the security status in the upper right corner.

10.3 Code Signing

Software manufacturers digitally sign their software distributions in order to authenticate their software and respective updates to the users, a practice also known as *code signing*. Code signing is a very important countermeasure against attacks that use viruses and Trojan horses.

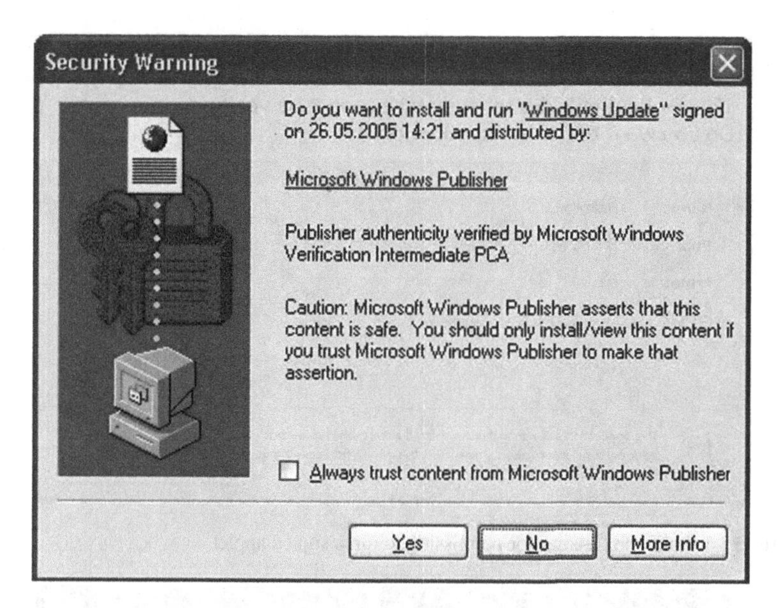

Fig. 10.14 Installing a digitally signed windows update

Signed software should only be installed if the signature has been successfully verified. Figure 10.14 shows the protection of Microsoft Windows updates. The same is done for most other operating systems. Although it is not secure, in some applications users may also choose to install software with unverified signatures.

Code signing can also be used to give certain rights to Java applets that run in a Java-enabled browser. For example, signed applets may be allowed to read from and write to the client file system, to establish connections to the Internet, or to communicate with external ports such as the serial or parallel port. Figures 10.15–10.17 show the windows that appear during signature verification.

Another application of code signing is WebStart [18], which allows distributing applications over a network. This works as follows. The WebStart server distributes the application. Clients connect to the server and download the WebStart application which then runs on the client machine just as applets run in a browser. The WebStart application checks for updates, which it downloads automatically. This mechanism simplifies bug fixing and adding new features. Code signing authorizes WebStart applications to access resources on the client computer.

In certificates that are being used for code signing, the extended key usage extension should be set and contain the `codeSigning` OID "1.3.6.1.5.5.7.3.3". Otherwise, most applications cannot use the certificates in this context.

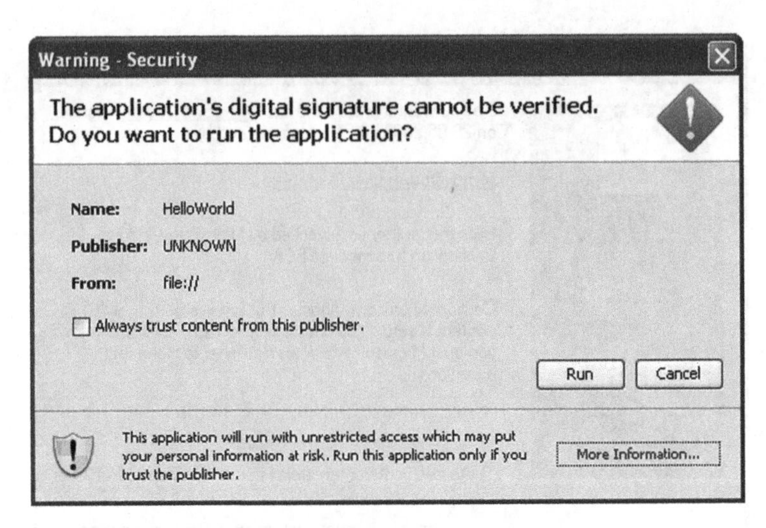

Fig. 10.15 A dialog box asking for permission to run a signed applet

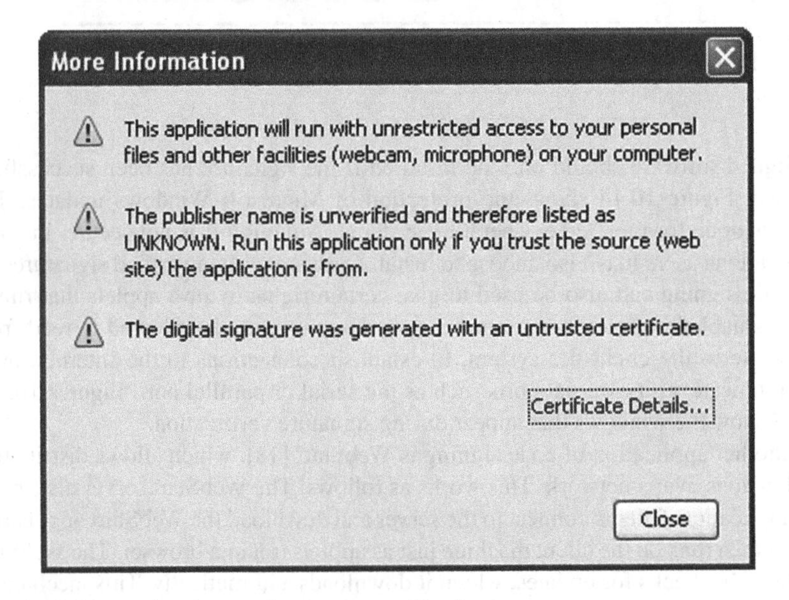

Fig. 10.16 Information about applet signature verification

10.4 VPN

PKI is also useful for protecting virtual private networks (VPNs). VPNs establish connections between remote computers and local area networks (LANs) as if they were part of the network. For example, VPNs are used by employees on business trips to connect to the LAN of their company or to connect branches of a company in

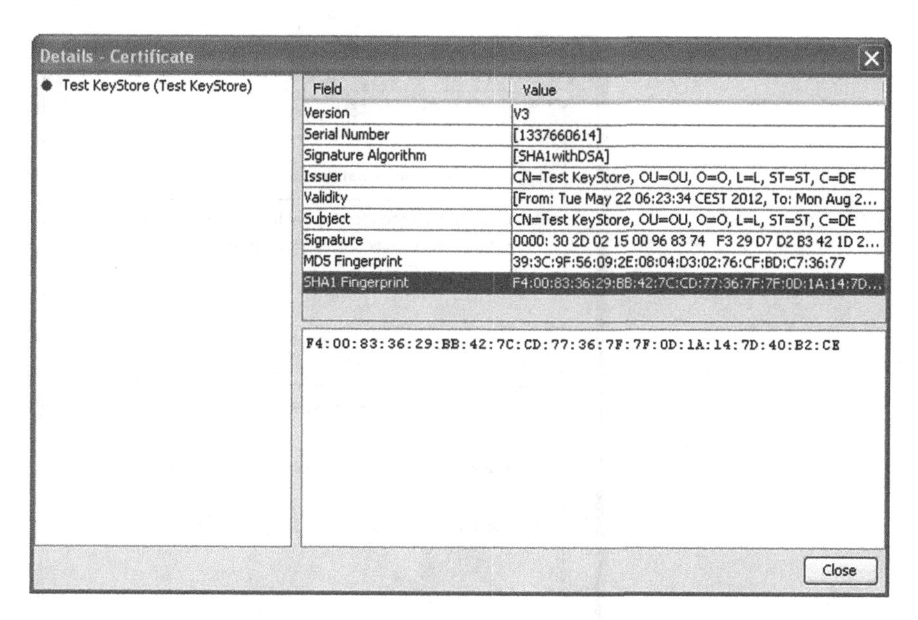

Fig. 10.17 Information about the certificate that is used to verify the applet signature

different countries. VPN security is implemented based on protocols such as IPsec [15], SSH [23], or TSL/SSL [5]. The frequently used VPN software OpenVPN is based on TSL/SSL and supports X.509 certificate-based authentication.

We show an example of how VPN is used. Alice remotely connects her laptop to the LAN of her university research group using OpenVPN. She uses her client certificate to authenticate herself to the VPN server. Conversely, the server authenticates itself to Alice using its server certificate. Figure 10.18 shows Alice's client certificate. Note that the certificate is not issued to Alice, but to her laptop `lap58`. Figure 10.19 shows the root certificate that is used by Alice's VPN client to authenticate the VPN server.

As seen in Figs. 10.20 and 10.21, both certificates are involved in establishing the VPN connection. Figure 10.20 shows Alice's VPN configuration file. Line 9 specifies the file name of the VPN root certificate. Lines 10 and 11 specify the file names of Alice's laptop's certificate and private key. Line 17 specifies the DN of the VPN server to connect to. Figure 10.21 shows the log file resulting from Alice establishing the VPN connection. Lines 288–290 document the successful verification of the VPN gateway certificate. Lines 291–294 show the so-called cipher suite that defines which cryptographic mechanisms are used for the TLS channel. They are `AES-128-CBC` and `SHA1` for both directions (to and from the server), which means that the data is encrypted using 128-bit AES in cipher block chaining (CBC) mode and authenticated with HMAC [16] using the SHA1 hash algorithm.

Fig. 10.18 Alice's VPN certificate

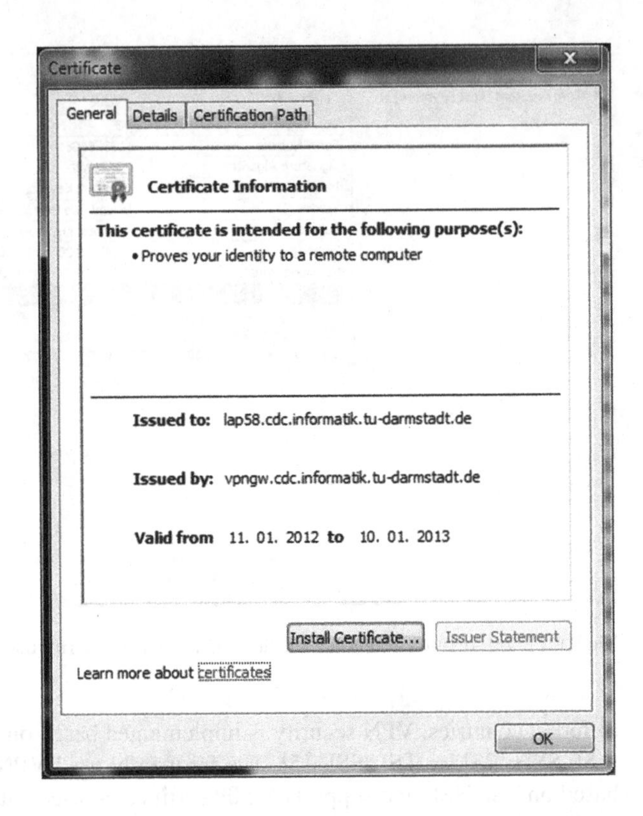

10.5 Legally Binding Electronic Signatures

Digital signatures may also replace handwritten signatures in legally binding contracts when the contracts were issued electronically. Legally binding signatures are defined in related signature laws. An overview of such laws of various countries can be found in [8].

Many European countries have laws that are compatible with the 1999/93/EC directive of the European Union [9]. This directive specifies the *advanced electronic signature*. Such a signature must be uniquely linked to the signer; it must support signer identification and protection against unauthorized modification of the signed data. The European directive requires certificates that are used for legally binding signatures, so-called *qualified certificates*, to be issued by certification service providers that must satisfy a number of requirements such as providing directory services, supporting immediate certificate revocation, and obtaining and verifying the identity of the certificate owner. If the certificate owner chooses to use a pseudonym, the CSP must be able to link the pseudonym with the identity of the certificate owner. The directive does not specify the certificate format, but X.509 is the defacto standard.

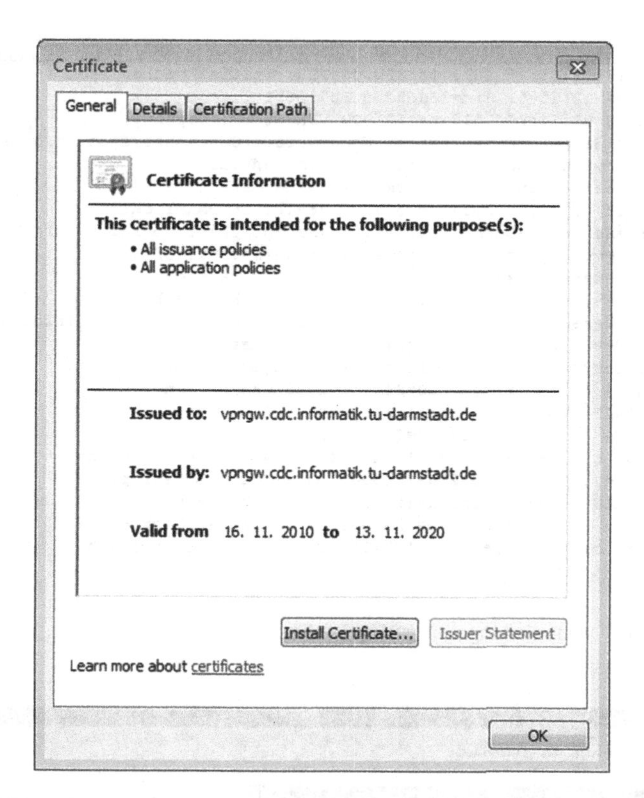

Fig. 10.19 VPN root certificate

```
1 # CDC-lap58-VPN Config
2 remote 130.83.167.12 1195
3 proto udp
4 client
5 dev tap
6 resolv-retry infinite
7 nobind
8 mute-replay-warnings
9 ca ca.crt
10 cert lap58.cdc.informatik.tu-darmstadt.de.crt
11 key lap58.cdc.informatik.tu-darmstadt.de.key
12 cipher AES-128-CBC
13 comp-lzo
14 verb 4
15 persist-tun
16 redirect-gateway def1 bypass-dhcp
17 tls-remote /C=DE/ST=Hessen/L=Darmstadt/O=TU_Darmstadt/
 . OU=Fachbereich_Informatik/OU=CDC/CN=vpngw-nat.cdc.informatik.tu-
 . darmstadt.de/emailAddress=admins@cdc.informatik.tu-darmstadt.de
```

Fig. 10.20 VPN configuration file

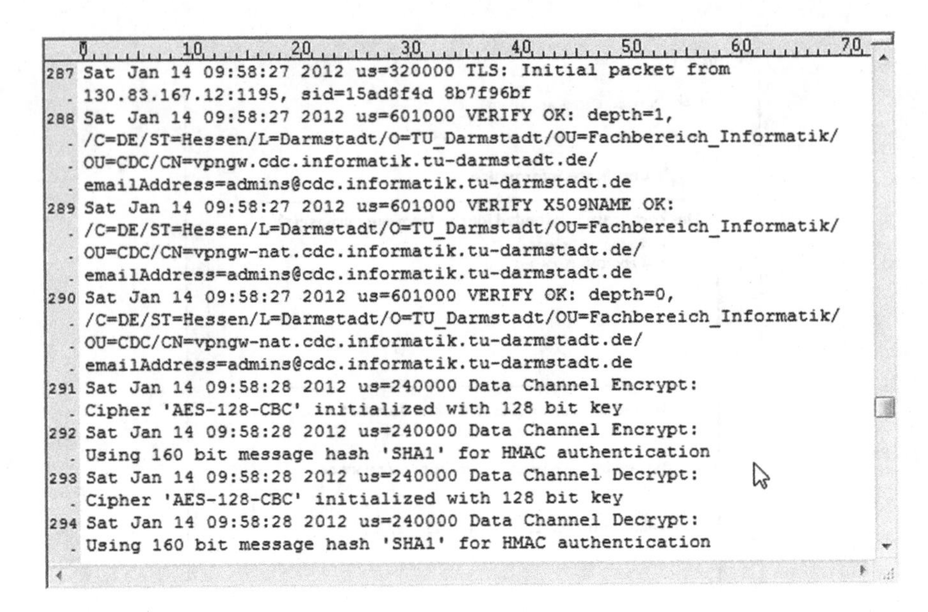

287 Sat Jan 14 09:58:27 2012 us=320000 TLS: Initial packet from 130.83.167.12:1195, sid=15ad8f4d 8b7f96bf
288 Sat Jan 14 09:58:27 2012 us=601000 VERIFY OK: depth=1, /C=DE/ST=Hessen/L=Darmstadt/O=TU_Darmstadt/OU=Fachbereich_Informatik/OU=CDC/CN=vpngw.cdc.informatik.tu-darmstadt.de/emailAddress=admins@cdc.informatik.tu-darmstadt.de
289 Sat Jan 14 09:58:27 2012 us=601000 VERIFY X509NAME OK: /C=DE/ST=Hessen/L=Darmstadt/O=TU_Darmstadt/OU=Fachbereich_Informatik/OU=CDC/CN=vpngw-nat.cdc.informatik.tu-darmstadt.de/emailAddress=admins@cdc.informatik.tu-darmstadt.de
290 Sat Jan 14 09:58:27 2012 us=601000 VERIFY OK: depth=0, /C=DE/ST=Hessen/L=Darmstadt/O=TU_Darmstadt/OU=Fachbereich_Informatik/OU=CDC/CN=vpngw-nat.cdc.informatik.tu-darmstadt.de/emailAddress=admins@cdc.informatik.tu-darmstadt.de
291 Sat Jan 14 09:58:28 2012 us=240000 Data Channel Encrypt: Cipher 'AES-128-CBC' initialized with 128 bit key
292 Sat Jan 14 09:58:28 2012 us=240000 Data Channel Encrypt: Using 160 bit message hash 'SHA1' for HMAC authentication
293 Sat Jan 14 09:58:28 2012 us=240000 Data Channel Decrypt: Cipher 'AES-128-CBC' initialized with 128 bit key
294 Sat Jan 14 09:58:28 2012 us=240000 Data Channel Decrypt: Using 160 bit message hash 'SHA1' for HMAC authentication

Fig. 10.21 VPN connection log

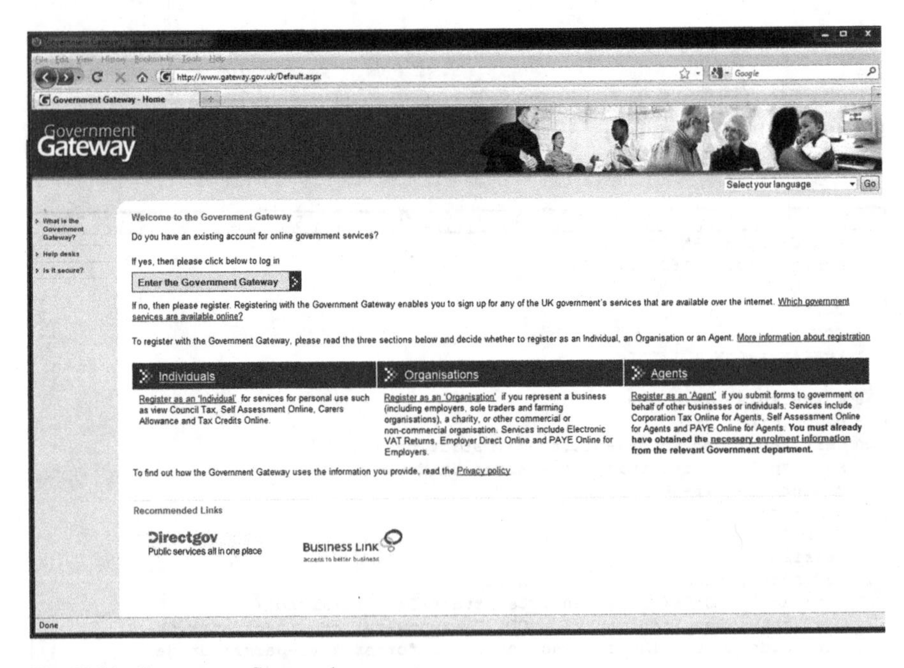

Fig. 10.22 Government Gateway homepage

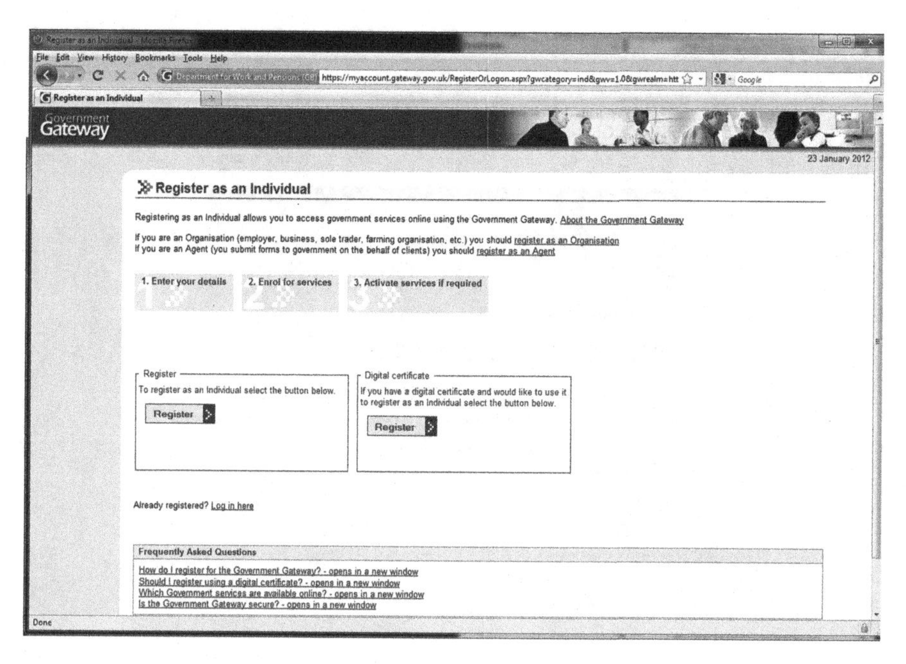

Fig. 10.23 Government Gateway registration

The German law also defines so-called *qualified signatures* that compared with advanced signatures have many additional features. For example, such signatures must be created using secure devices specified by the German signature ordinance.

In Germany, qualified signatures may be used in many contexts as a replacement of handwritten signatures. We present a few examples. Certain companies must document their waste management. Such documentation requires legally binding signatures which may be issued as qualified signatures on electronic documents. Electronic documentation allows for very efficient management. Another application of qualified signatures is e-justice, for example interaction with the courts over the Internet. Also, the German patent office accepts applications that are signed with qualified signatures.

10.6 E-Government

PKI is also useful to secure e-government applications. This can be done using legally binding electronic signatures as described in Sect. 10.5, but also by other mechanisms. The term e-government refers to electronic transactions between a government and citizens, companies, or other entities within this government's range of authority. Frequently, strong authentication, protection of integrity and confidentiality is a legal requirement for such transactions. Examples of e-government

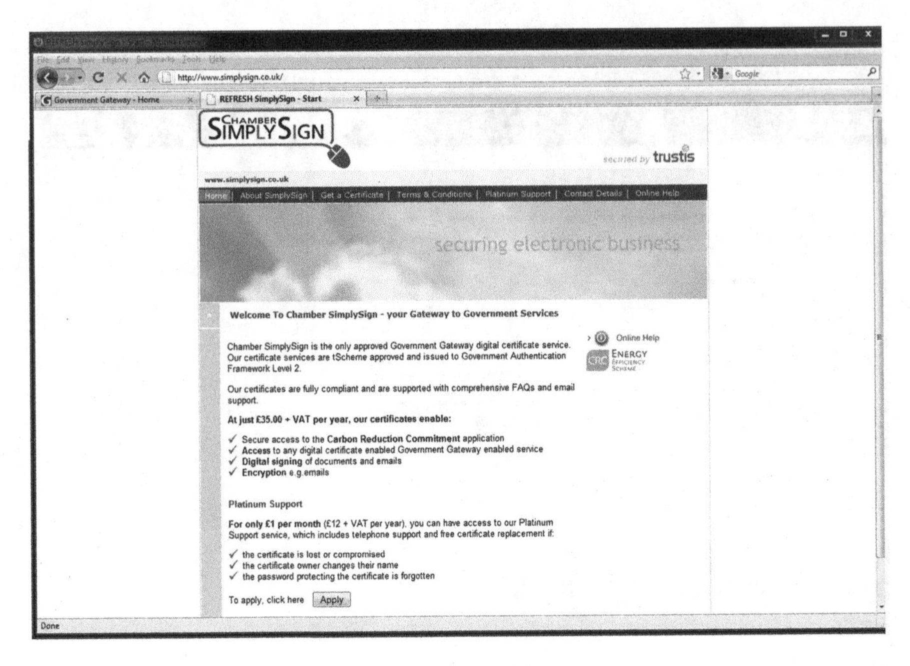

Fig. 10.24 SimplySign homepage

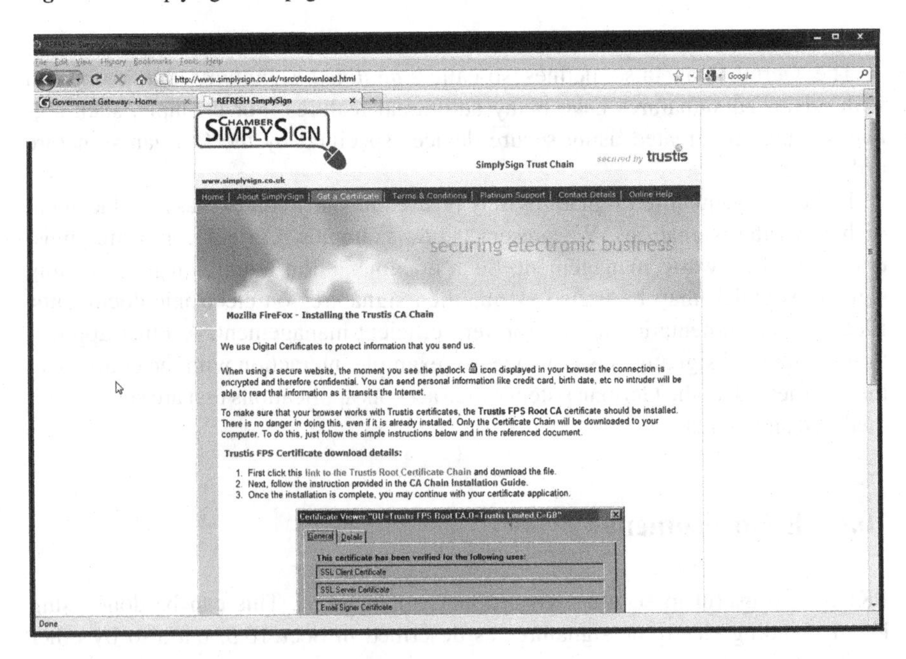

Fig. 10.25 SimplySign certification path download

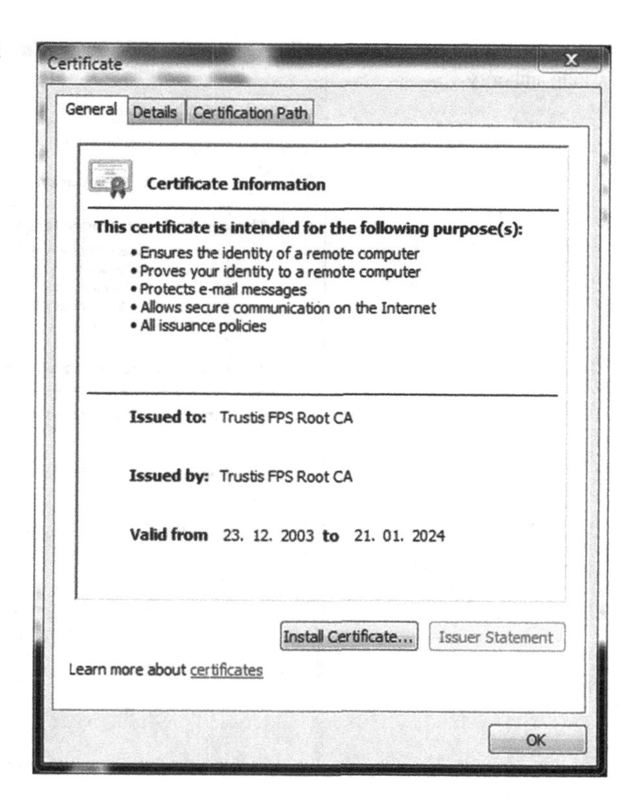

Fig. 10.26 SimplySign root certificate

applications are electronic citizen portals, electronic voting, and electronic tax declarations.

For example, if the London citizen Alice wants to manage her council tax account online, she first has to register with the service. Being a UK citizen, she does this at the UK Government Gateway [3]. As seen in Fig. 10.22, this gateway allows for registration of individuals, organizations and agents that act on behalf of others. She follows the registration process for individuals. In the first step depicted in Fig. 10.23 she has to decide whether she wants to use a certificate or not.

Alice is security-aware, and decides to use a certificate. As she does not yet have a proper certificate, she has to apply for one from a proper CSP. In our example, Alice decides to get a certificate from SimplySign [22]. Figure 10.24 shows SimplySign's home page, where we can see that Alice has to pay an annual fee for her certificate. She accepts the registration with SimplySign. As seen in Fig. 10.25, the first thing she has to do is to download and install the respective certification path to her browser.

Figure 10.26 shows SimplySign's root certificate, which is a self-signed certificate. Figure 10.27 shows SimplySign's issuing authority certificate, which is signed by SimplySign's root.

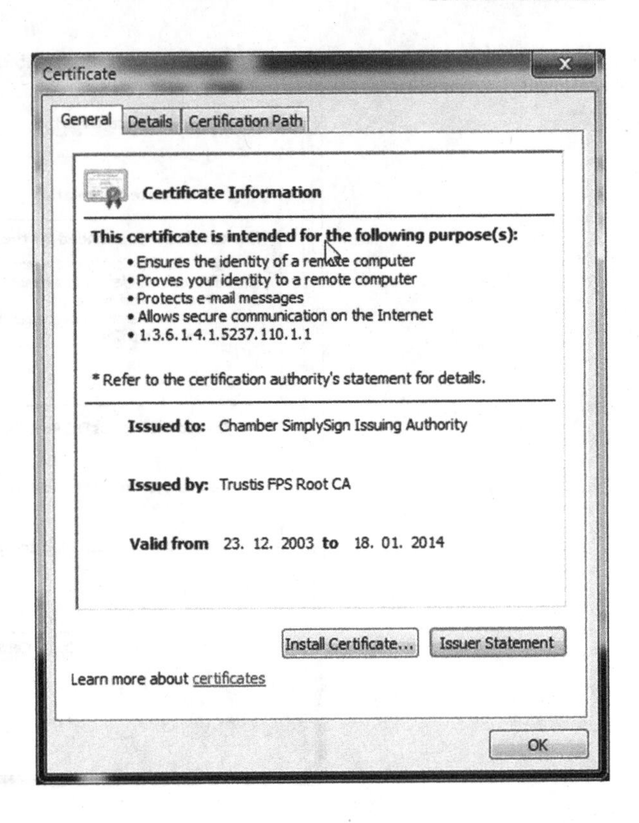

Fig. 10.27 SimplySign issuing authority

After applying for her certificate and installing it in her browser, she goes back to the Web site of the Government Gateway and finalizes the registration using her certificate. Once logged in, she can choose the e-government service she wants to use from a list. Figure 10.28 shows the list as presented by the Government Gateway. Note that for each service the provision of further information, such as the tax number, and additional registration might be necessary.

10.7 Exercises

10.1. Alice wants to send an encrypted email to Bob. Although she possesses Bob's certificate and has installed it in her email client, the client refuses to send the email. Give possible reasons for the behavior of the email client.

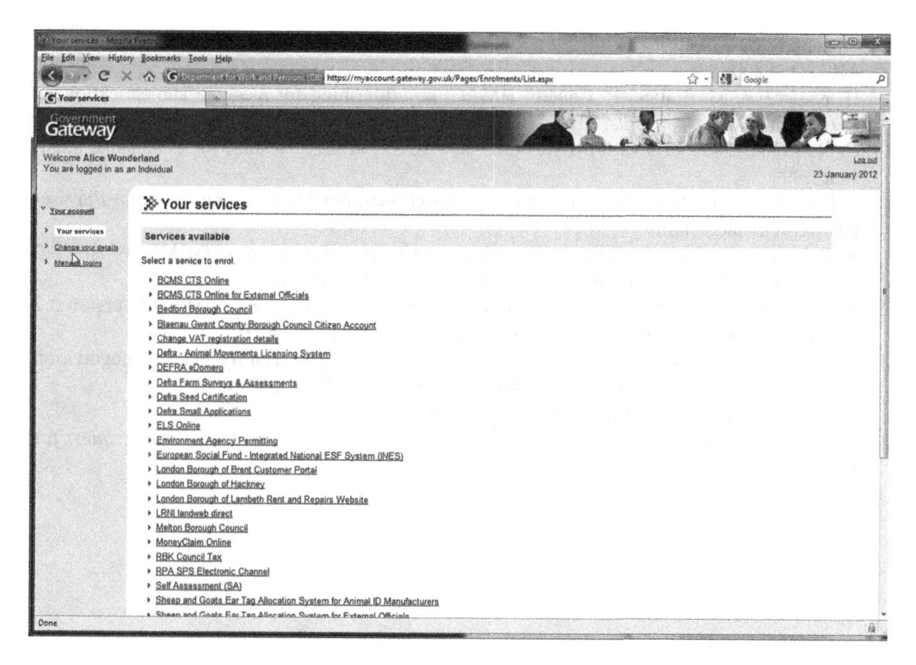

Fig. 10.28 Government Gateway services

References

1. J. Callas, L. Donnerhacke, H. Finney, D. Shaw, R. Thayer, OpenPGP message format, in *IETF Request for Comments*, 4880, Nov 2007
2. D. Cooper, S. Santesson, S. Farrell, S. Boeyen, R. Housley, W. Polk, Internet X.509 public key infrastructure certificate and certificate revocation list (CRL) profile, in *IETF Request for Comments*, 5280, May 2008
3. Department for Work and Pensions: Government Gateway, http://www.gateway.gov.uk/
4. T. Dierks, E. Rescorla, The transport layer security (TLS) protocol version 1.1, in *IETF Request for Comments*, 4346, Apr 2006
5. T. Dierks, E. Rescorla, The transport layer security (TLS) protocol version 1.2, in *IETF Request for Comments*, 5246, Aug 2008
6. M. Elkins, D. Del Torto, R. Levien, T. Roessler, MIME security with OpenPGP, in *IETF Request for Comments*, 3156, Aug 2001
7. EnigMail project, http://whttp://www.enigmail.net
8. eSignatures Legal Wiki, http://www.esignaturelegalwiki.org
9. European Parliament and Council, Directive 1999/93/EC of the European Parliament and of the Council of 13 Dec 1999 on a Community framework for electronic signatures, Dec 1999
10. P. Ford-Hutchinson, Securing FTP with TLS, in *IETF Request for Comments*, 4217, Oct 2005
11. A. Freier, P. Karlton, P. Kocher, The secure sockets layer (SSL) protocol version 3.0, in *IETF Request for Comments*, 6101, Aug 2011
12. GNU Privacy Guard, http://www.gnupg.org/
13. R. Harrison, Lightweight directory access protocol (LDAP): authentication methods and security mechanisms, in *IETF Request for Comments*, 4513, June 2006
14. P. Hoffman, SMTP service extension for secure SMTP over transport layer security, in *IETF Request for Comments*, 3207, Feb 2002

15. S. Kent, K. Seo, Security architecture for the internet protocol, in *IETF Request for Comments*, 4301, Dec 2005
16. H. Krawczyk, M. Bellare, R. Canetti, HMAC: keyed-hashing for message authentication, in *IETF Request for Comments*, 2104, Feb 1997
17. C. Newman, Using TLS with IMAP, POP3 and ACAP, in *IETF Request for Comments*, 2595, June 1999
18. ORACLE, Java Web Start Technology, http://www.oracle.com/technetwork/java/javase/javawebstart/index.html
19. B. Ramsdell, S. Turner, Secure/multipurpose internet mail extensions (S/MIME) version 3.2 certificate handling, in *IETF Request for Comments*, 5750, Jan 2010
20. B. Ramsdell, S. Turner, Secure/multipurpose internet mail extensions (S/MIME) version 3.2 message specification, in *IETF Request for Comments*, 5751, Jan 2010
21. E. Rescorla, M. Ray, S. Dispensa, N. Oskov, Transport layer security (TLS) renegotiation indication extension, in *IETF Request for Comments*, 5746, Feb 2010
22. SimplySign, SimplySign, http://www.simplysign.co.uk/
23. T. Ylonen, C. Lonvick, The secure shell (SSH) protocol architecture, in *IETF Request for Comments*, 4251, Jan 2006

Appendix A
Basic Path Validation Algorithm

The variables that are used in the algorithm are listed in Table A.1

The algorithm builds the so-called `valid_policy_tree`, which is used to decide whether there is a valid policy for a certificate path. The answer is "yes" if the algorithm is able to construct a valid policy tree of depth n. Otherwise, the algorithm constructs a null policy tree. The nodes of this tree are called *policy nodes*. As shown in Fig. A.1, each policy node has three attributes. The first attribute is the `valid_policy`. It is one of the policies from the certificate being processed by the algorithm. The second attribute is the `qualifier_set`, which contains qualifiers of this policy. The third attribute is the `expected_policy_set`. One of the policies in this set must be valid for the next certificate in the chain in order for the tree construction to continue. We represent a policy node by $N = \langle vp, \{qs\}, \{eps\} \rangle$ where N represents the node, vp is the valid_policy, qs the qualifier_set, and eps the expected_policy_set.

J.A. Buchmann et al., *Introduction to Public Key Infrastructures*,
DOI 10.1007/978-3-642-40657-7,

Table A.1 Variables of RFC 5280 path validation algorithm

cp	:	certification path
n	:	length of *cp*
i	:	iteration counter
C	:	current certificate
uips	:	user-initial-policy-set
ipolmap_inh	:	initial-policy-mapping-inhibit
iexpol	:	initial-explicit-policy
iapol_inh	:	initial-any-policy-inhibit
ipersub	:	initial-permitted-subtrees
iexsub	:	initial-excluded-subtrees
v_p_t	:	valid_policy_tree
p_st	:	permitted_subtrees
ex_st	:	excluded_subtrees
ex_pol	:	explicit_policy
in_ap	:	inhibit_any-policy
pol_map	:	policy_mapping
w_pk	:	working_public_key
w_pk_a	:	working_public_key_algorithm
w_pk_p	:	working_public_key_parameters
w_iss	:	working_issuer_name
m_path	:	max_path_length
ex_pol_s	:	expected_policy_set
v_p_n_s	:	valid_policy_node_set

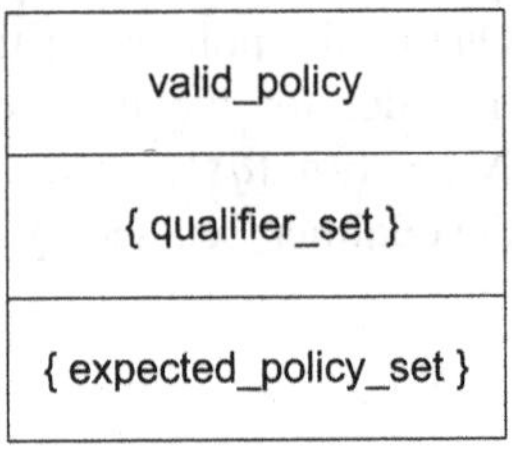

Fig. A.1 Policy node

Algorithm 1 Initialization

Input: *cp* of length *n*
Input: *date* = current time
Input: *uips*
Input: Trust anchor information like its name, public key along with its algorithm and parameters
Input: *ipolmap_inh*
Input: *iexpol*
Input: *iapol_inh*
Input: *ipersub*
Input: *iexsub*

1: $v\_p\_t = \langle anyPolicy, \{\}, \{anyPolicy\}\rangle$
2: $p\_st = ipersub$
3: $ex\_st = iexsub$
4: **if** $iexpol = true$ **then** $ex\_pol = 0$ **else** $ex\_pol = n + 1$ **end if**
5: **if** $iapol\_inh = true$ **then** $in\_ap = 0$ **else** $in\_ap = n + 1$ **end if**
6: **if** $ipolmap\_inh = true$ **then** $pol\_map = 0$ **else** $pol\_map = n + 1$ **end if**
7: set $w\_pk\_a, w\_pk, w\_pk\_p,$ and w_iss according to trust anchor information
8: m_path = n
9: C = first certificate in the path
10: $i = 0$

Algorithm 2 Basic certificate processing

```
1: i = i + 1
2: if C is not signed with w_pk_a, w_pk, and w_pk_p then
3:     return failure and reason
4: end if
5: if C's notBefore < date < C's notAfter then
6:     return failure and reason
7: end if
8: if C is revoked then
9:     return failure and reason
10: end if
11: if C's issuerDN ≠ w_iss_name then
12:     return failure and reason
13: end if
14: if C is self-issued and it is not the last certificate in cp then
15:     goto 21
16: else
17:     if C'c subject DN and every subject alternative name contained in C is not contained in
        p_st or is contained in ex_st then
18:         return failure and reason
19:     end if
20: end if
21: if C has the certificate policies extension then
22:     if v_p_t is NOT NULL then
23:         process Policies (Algorithm 3)
24:     end if
25: else
26:     v_p_t = NULL
27: end if
28: if ex_pol > 0 OR v_p_t is NOT NULL then
29:     continue
30: else
31:     return failure and reason
32: end if
```

Algorithm 3 Process policies

1: **for** each policy in C other than anyPolicy **do**
2: p_oid = current policy
3: p_q = current policy qualifier
4: **for** each node N in v_p_t of depth $i-1$ with p_oid in ex_pol_s of N **do**
5: append child node $N' = \langle p\_oid, \{p\_q\}, \{p\_oid\}\rangle$
6: **if** node was not appended in previous step **then**
7: **for** each node N in v_p_t of depth $i-1$ with anyPolicy as $valid\_policy$ **do**
8: append child node $N' = \langle p\_oid, \{p\_q\}, \{p\_oid\}\rangle$
9: **end for**
10: **end if**
11: **end for**
12: **end for**
13: **if** C contains the policy anyPolicy **then**
14: p_q = policy qualifier of anyPolicy
15: **if** $in\_ap > 0$ OR ($i < n$ AND C is self-issued) **then**
16: **for** each node N in v_p_t of depth $i-1$ **do**
17: **for** each policy q_oid in ex_pol_s of N where q_oid does not appear as a child of N **do**
18: append child node $N' = \langle q\_oid, \{p\_q\}, \{q\_oid\}\rangle$
19: **end for**
20: **end for**
21: **end if**
22: **end if**
23: starting from $i-1$ recursively delete all nodes of depth less or equal to $i-1$ that have no child nodes

Algorithm 4 Preparation for certificate $i + 1$

1: process policy mappings (Algorithm 5)
2: *w_iss* = subjectDN of *C*
3: *w_pk* = subjectPublicKey of *C*
4: **if** *C* has subjectPublicKey parameters **then**
5:     *w_pk_p* = subjectPublicKey parameters of *C*
6: **else**
7:     **if** $w\_pk\_a \neq$ subjectPublicKey algorithm of *C* **then**
8:         $w\_pk\_p = null$
9:     **end if**
10: **end if**
11: *w_pk_a* = subjectPublicKey algorithm of *C*
12: **if** *C* has the name constraints extension **then**
13:     **if** *C* contains permittedSubtrees **then**
14:         $p\_st = p\_st \cap$ permittedSubtrees of *C*
15:     **end if**
16:     **if** *C* contains excludedSubtrees **then**
17:         $ex\_st = ex\_st \cup$ excludedSubtrees of *C*
18:     **end if**
19: **end if**
20: **if** *C* is not self-issued **then**
21:     $ex\_pol = \max(0, ex\_pol - 1)$
22:     $pol\_map = \max(0, pol\_map - 1)$
23:     $in\_ap = \max(0, in\_ap - 1)$
24: **end if**
25: **if** requireExplicitPolicy is set in *C* **then**
26:     *ex_pol* = min(*ex_pol*, requireExplicitPolicy of *C*)
27: **end if**
28: **if** inhibitPolicyMapping is set in *C* **then**
29:     *pol_map* = min(*pol_map*, inhibitPolicyMapping of *C*)
30: **end if**
31: **if** inhibitAnyPolicy is set in *C* **then**
32:     *in_ap* = min(*in_ap*, inhibitAnyPolicy of *C*)
33: **end if**
34: **if** *C* is not a CA certificate **then**
35:     **return** failure and reason
36: **end if**
37: **if** *C* is not self-issued **then**
38:     **if** $m\_path \leq 0$ **then**
39:         **return** failure and reason
40:     **else**
41:         $m\_path = m\_path - 1$
42:     **end if**
43: **end if**
44: **if** pathLenContraint is set in *C* **then**
45:     *m_path* = min(*m_path*, pathLenConstraints of *C*)
46: **end if**
47: **if** *C* has the key usage extension **then**
48:     **if** keyCertSign bit is not set **then**
49:         **return** failure and reason
50:     **end if**
51: **end if**
52: recognize and process other extensions of *C* and **return** failure and reason if an unrecognized critical extension is found
53: *C* = next certificate in path

Algorithm 5 Process policy mappings

1: **if** C has the policy mappings extension **then**
2: **if** *anyPolicy* is found in issuerDomainPolicy or subjectDomainPolicy **then return** failure and reason
3: **end if**
4: **for** each issuerDomainPolicy **do**
5: id_p = current issuerDomainPolicy
6: $spol$ = set of policies that id_p is mapped to
7: **if** $pol\_map > 0$ **then**
8: $match\_found$ = false
9: **for** each node N in v_p_t of depth i with $valid\_policy = id\_p$ **do**
10: ex_pol_s of $N = spol$
11: $match\_found$ = true
12: **end for**
13: **if** $match\_found$ = false but there is a node N' of depth i with $valid\_policy$ = *anyPolicy* **then**
14: p_q = policy qualifier of node N'
15: append a node Q to the father of N' with $Q = \langle id\_p, \{p\_q\}, \{spol\}\rangle$
16: **end if**
17: **else**
18: delete each node N of depth i with $valid\_policy = id\_p$
19: starting from $i - 1$ recursively delete all nodes of depth less or equal to $i - 1$ that have no child nodes
20: **end if**
21: **end for**
22: **else**
23: **return**
24: **end if**

Algorithm 6 Wrap-up procedure

1: $ex\_pol = \min(0, ex\_pol - 1)$
2: **if** C has requireExplicitPolicy = 0 **then**
3: $ex\_pol = 0$
4: **end if**
5: w_pk = subjectPublicKey of C
6: **if** C has subjectPublicKey parameters **then**
7: w_pk_p = subjectPublicKey parameters of C
8: **else**
9: **if** $w\_pk\_a \neq$ subjectPublicKey algorithm of C **then**
10: $w\_pk\_p = null$
11: **end if**
12: **end if**
13: w_pk_a = subjectPublicKey algorithm of C
14: recognize and process other extensions of C and **return** failure and reason if an unrecognized critical extension is found
15: **if** $v\_p\_t \neq$ NULL **then**
16: **if** *uips* is not anyPolicy **then**
17: $v\_p\_n\_s$ = set of nodes whose parent nodes have *valid_policy* anyPolicy
18: **for** each node N in $v\_p\_n\_s$ **do**
19: **if** *valid_policy* of N is not in the *uips* and is not anyPolicy **then**
20: delete N and its children nodes
21: **end if**
22: **end for**
23: **if** v_p_t includes a node N of depth n with *valid_policy* anyPolicy and $uips \neq$ anyPolicy **then**
24: p_q = qualifier set of node N
25: **for** each p_oid in $uips$ that is not the $valid\_policy$ of a node N in the $v\_p\_n\_s$ **do**
26: append a node N' to the father node of N with $N' = (p\_oid, \{p\_q\}, \{p\_oid\})$ and delete N
27: starting from $i - 1$ recursively delete all nodes of depth less or equal to $i - 1$ that have no child nodes
28: **end for**
29: **end if**
30: **else**
31: goto 36
32: **end if**
33: **else**
34: v_p_t = NULL
35: **end if**
36: **if** $ex\_pol > 0$ OR $v\_p\_t \neq$ NULL **then**
37: **return** success and output v_p_t, w_pk, w_pk_a, w_pk_p
38: **else**
39: **return** failure and reason
40: **end if**

Solutions to the Exercises

Exercises of Chap. 1

1.1 The dealer is interested in maintaining the integrity of the list of books. The messages of the dealer and the customer need to be authenticated. The transactions need to be confidential to protect the privacy of the customers and the financial information. The order and reply of the dealer must be binding.

1.2 Answer:

- Properties: Fingerprint, iris, face, or other biometric properties.
- Abilities: Keyboard stroke, voice.
- Knowledge: PIN, name of pet.
- Possession: Smart card, credit card, key.

1.3 Answer:

1. Accessing the database does not enable an adversary to impersonate users since cryptographic hash functions are one-way. Replay is possible by intercepting the password hashes and reusing them at a later point in time.
2. The replay attack can be prevented by using a random nonce n as shown in Fig. A.2.

   The nonce may also be the current time.

Exercises of Chap. 2

2.1 There are three versions of X.509 certificates. X509v1 certificates do not contain the issuerUniqueID, the subjectUniqueID, or extensions. X509v2 certificates contain the issuerUniqueID and subjectUniqueID fields but no extensions. X509v3 certificates contain extensions and these two fields.

2.2 The certificates are as shown in Figs. A.3 and A.4:

DOI 10.1007/978-3-642-40657-7, © Springer-Verlag Berlin Heidelberg 2013

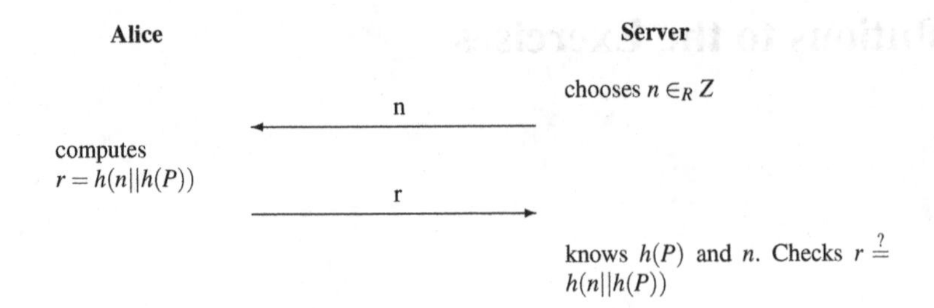

Fig. A.2 Preventing replay attack using nonce

Certificate 1	
Serial No.:	26540
Issuer:	CN=Test CA
NotBefore:	2004-04-03
NotAfter:	2005-04-03
Subject:	CN=Alice
Public Key:	key-0x4D367AB9
X509v3Extensions:	
KeyUsage: critical digitalSignature, dataEnciphernemt	
Subject Key Identifier: keyId: 12:AB:45:76:F8:98	
Authority Key Identifier: keyId: BE:76:34:4E:60:34 aci: CN=Master CA acsn: 34	

Certificate 2	
Serial No.:	1
Issuer:	CN=Master CA
NotBefore:	2003-11-14
NotAfter:	2008-11-14
Subject:	CN=Master CA
Public Key:	key-0x18FF6542
X509v3Extensions:	
Basic Constraints: critical CA: TRUE pathlen: 1	
KeyUsage: critical keyCertSign, cRLSign	
Subject Key Identifier: keyId: 11:23:34:AB:65:F0	
Authority Key Identifier: keyId: 11:23:34:AB:65:F0 aci: CN=Master CA acsn: 1	

Fig. A.3 Certificates 1 and 2 with AKI values

Certificate 3	
Serial No.:	34
Issuer:	CN=Master CA
NotBefore:	2003-11-15
NotAfter:	2008-11-10
Subject:	CN=Test CA
Public Key:	key-0x347893B2
X509v3Extensions:	
KeyUsage: critical keyCertSign, cRLSign	
Subject Key Identifier: keyId: BE:76:34:4E:60:34	
Authority Key Identifier: keyId: 11:23:34:AB:65:F0 aci: CN=Master CA acsn: 1	

Fig. A.4 Certificate 3 with AKI values

2.3 Answer:

1. Yes, in self-signed certificates.
2. Yes, if an issuer has issued two certificates and signed them with the same key.
3. Yes, for certificates that certify the same public key, for example, due to certificate renewal when a certificate expires.

2.4 Following the ASN.1 definition of the key usage extension in Listing 2.6, the key contained in the certificate can be used for digital signatures, non-repudiation, and data encipherment. For example, this is possible for RSA keys.

2.5 This is permitted because the certificates have been issued by different issuers as indicated by the issuer field.

Exercises of Chap. 3

3.1 Table A.2 shows Alice's key ring with all values.

3.2 Answer:

1. The hash value of a complete X.509 certificate is used for calculating the fingerprint of the certificate. This fingerprint allows us to verify the integrity of the certificate, for example, in the context of direct trust.
2. The hash value of the TBS part of an X.509 certificate is used for creating and verifying the signature of the certificate.

3.3 Answer:

1. Alice may trust in any of the three CAs: Root, Org1, or Dep1.
2. Alice may trust in any of the CAs: Root, Org1, or Dep2.
3. Yes. Alice may trust in the Org2 CA directly.
4. No. Alice may trust in the Root CA directly. This would lead to indirect trust in Org2. Another possibility would be to trust in *Org2* directly, which is not desired.

3.4 Answer:

1. Trusted Lists: as entities A, E, H, M, and O are the trust anchors of every end entity, it is enough to add them to each end entity's trusted list.

    Bridge: the individual PKIs can be bridged as seen in Fig. A.5.

    Cross-certification: the individual PKIs can be combined by cross-certification as shown in Fig. A.6.

    Common Root: the individual PKIs can be combined by putting them under a common root CA as shown in Fig. A.7.

Table A.2 Public key ring of Alice with all values

Public key ring of **Alice**			
1	Public key owner		: Alice
	Owner trust/key legitimacy		: ultimate/complete
	1	Signer/trust in signer	: Alice/ultimate
2	Public key owner		: Bob
	Owner trust/key legitimacy		: marginal/complete
	1	Signer/trust in signer	: Alice/ultimate
	2	Signer/trust in signer	: Bob/marginal
	3	Signer/trust in signer	: Carl/none
3	Public key owner		: Carl
	Owner trust/key legitimacy		: none/marginal
	1	Signer/trust in signer	: Bob/marginal
	2	Signer/trust in signer	: Carl/none
4	Public key owner		: Diana
	Owner trust/key legitimacy		: unknown/complete
	1	Signer/trust in signer	: Bob/marginal
	2	Signer/trust in signer	: Carl/none
	3	Signer/trust in signer	: Diana/unknown
	4	Signer/trust in signer	: Emil/marginal
	5	Signer/trust in signer	: Frank/marginal
5	Public key owner		: Emil
	Owner trust/key legitimacy		: marginal/complete
	1	Signer/trust in signer	: Alice/ultimate
	2	Signer/trust in signer	: Bob/marginal
	3	Signer/trust in signer	: Emil/marginal
6	Public key owner		: Frank
	Owner trust/key legitimacy		: marginal/complete
	1	Signer/trust in signer	: Alice/ultimate
	2	Signer/trust in signer	: Frank/marginal

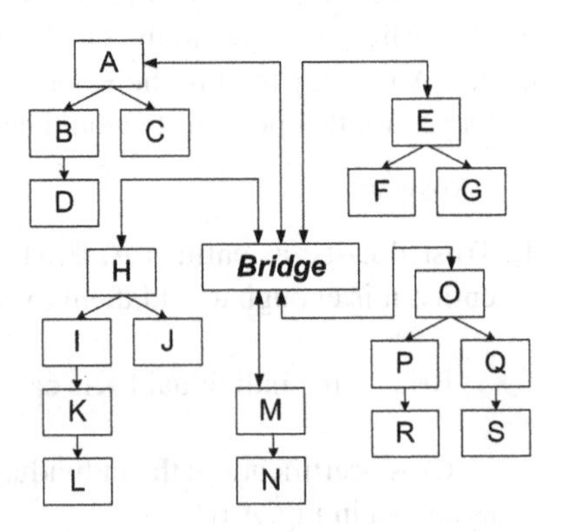

Fig. A.5 Bridged individual PKIs

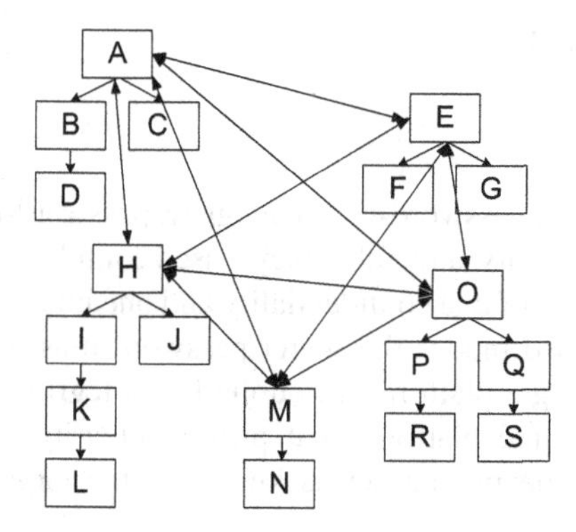

Fig. A.6 Cross-certified individual PKIs

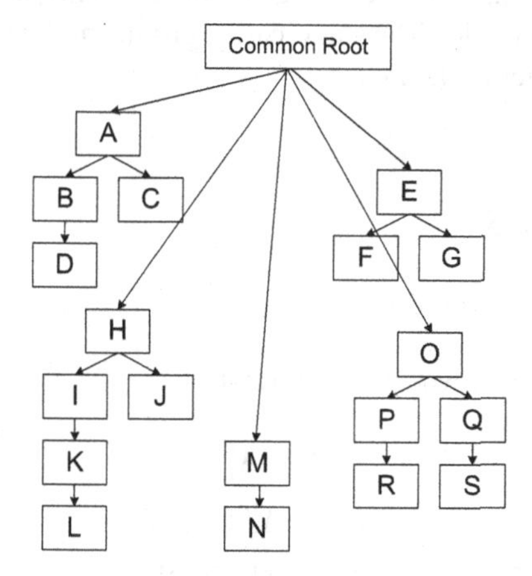

Fig. A.7 Individual PKIs under common root CA

Table A.3 Trust anchors

Architecture	Number of trust anchors	Trust anchor of G
Trusted lists	5	A, E, H, M, O
Bridge	1	E
Cross-certification	1	E
Common root	1	Common root

2. When trusted lists are used it is not necessary to issue any new certificate. In the bridge solution $2n = 10$ new certificates are issued. Cross-certification requires issuing $n * (n - 1) = 20$ new certificates. In the common root solution, $n = 5$ new certificates are issued.
3. Table A.3 shows the number of trust anchors required in each case and the trust anchors of entity G.

Exercises of Chap. 4

4.1 Answer:

1. There are four modes: password-based integrity, password-based confidentiality, public key-based integrity, and public key-based confidentiality.
2. All four combinations of one confidentiality and one integrity mode are possible.
3. In the password-based modes the secret password must be known both to the source and to the target platform. In public key integrity mode the private key is known to the source platform that proves integrity by signing. In public key confidentiality the private key is known at the target platform that must decrypt.

4.2 The hash of the document is computed on the PC. This hash is sent to the smart card via the smart card reader. The smart card signs the hash and sends the signature via the smart card reader to the PC.

Exercises of Chap. 5

5.1 Answer:

1. The user has complete revocation information.

$$F^1 + \Delta_1^5 \vdash F^5 \tag{A.1}$$

$$F^5 + \Delta_5^{10} \vdash F^{10} \tag{A.2}$$

$$F^{10} + \Delta_{10}^{15} \vdash F^{15} \tag{A.3}$$

$$F^{15} \vdash F^{14} \tag{A.4}$$

2. The user has complete revocation information.

$$F^{12} + \Delta_{10}^{14} \vdash F^{14} \tag{A.5}$$

3. The user does not have complete revocation information. F^2 is missing. If the user has F^2, then:

$$F^2 + \Delta_2^5 \vdash F^5 \tag{A.6}$$

$$F^5 + \Delta_5^{10} \vdash F^{10} \tag{A.7}$$

$$F^{10} + \Delta_{15}^{10} \vdash F^{15} \tag{A.8}$$

Table A.4 Revocation lists

CRL	CRL number	Base-CRL number
Full CRL on 01-01	0	N/A
Delta CRL on 01-15	2	0
Full CRL on 02-01	4	N/A
Delta CRL on 02-01	4	0
Delta CRL on 02-08	5	4
Delta CRL on 03-08	9	8

5.2 Answer:

1. See Table A.4.
2. The CRL issued on 1 March. It contains 78 entries. This is calculated by adding the number of all revoked certificates until this point in time, which is $12 + 4 + 6 + 7 + 1 + 9 + 14 + 3 + 22 = 78$.
3. The delta CRL issued on 1 March. It contains 48 entries. This is calculated by adding the number of all certificates revoked between 1 February and 1 March, which is $9 + 14 + 3 + 22 = 48$.
4. The user needs at least the complete CRL issued on 1 March. No delta CRLs are necessary. If the user has downloaded the complete CRL issued on 1 January, then the delta CRLs issued on 1 February and 1 March are necessary. If the user has downloaded the complete CRL issued on 1 February, then the delta CRL of 1 March is required.

5.3 Answer:

1. We have the following requirements:

$$\text{thisUpdate}_{CRL\,A} > \text{thisUpdate}_{CRL\,C}$$

because CRL A has a greater CRL number than CRL C.

$$\text{thisUpdate}_{CRL\,A} = \text{thisUpdate}_{CRL\,B}$$

because CRL A has the same CRL number as CRL B.

$$\text{thisUpdate}_{CRL\,A} < \text{nextUpdate}_{CRL\,A}$$
$$\text{thisUpdate}_{CRL\,B} < \text{nextUpdate}_{CRL\,B}$$
$$\text{thisUpdate}_{CRL\,C} < \text{nextUpdate}_{CRL\,C}$$

This implies the constellation as shown in Fig. A.8.

2. Yes, this can happen if more certificates than those contained in the Base CRL are revoked before the next complete CRL is issued.

CRL A	
Issuer:	CA1
ThisUpdate:	2011-08-15
NextUpdate:	2011-09-10
Revoked Certificates:	
Serial No.:	32
Serial No.:	16
Serial No.:	64
Serial No.:	128
X509v2 CRL Extensions:	
CRL Number:	non-critical 143

CRL B	
Issuer:	CA1
ThisUpdate:	2011-08-15
NextUpdate:	2011-09-10
Revoked Certificates:	
Serial No.:	64
Serial No.:	128
X509v2 CRL Extensions:	
CRL Number:	non-critical 143
Delta CRL Indicator: Base CRL Number:	critical 138

CRL C	
Issuer:	CA1
ThisUpdate:	2011-08-10
NextUpdate:	2011-08-15 or 2011-09-10
Revoked Certificates:	
Serial No.:	32
Serial No.:	16
X509v2 CRL Extensions:	
CRL Number:	non-critical 138

Fig. A.8 CRLs with valid values

CRL D	
Issuer:	CA1
ThisUpdate:	2011-10-15
NextUpdate:	2011-11-01
Revoked Certificates:	
Serial No.:	456
X509v2 CRL Extensions:	
CRL Number:	non-critical 321
Delta CRL Indicator: Base CRL Number:	critical 234

CRL E	
Issuer:	CA1
ThisUpdate:	2011-10-15
NextUpdate:	2011-11-15
Revoked Certificates:	
Serial No.:	232
Serial No.:	136
Serial No.:	164
Serial No.:	987
Serial No.:	456
X509v2 CRL Extensions:	
CRL Number:	non-critical 321

Fig. A.9 CRL number and base CRL number values

3. The CRLs are shown in Fig. A.9.
   The following combinations are permitted.

   (a) CRL $\text{Number}_{CRL\ D} = 321$
   Base CRL $\text{Number}_{CRL\ D} = 234$
   CRL $\text{Number}_{CRL\ E} = 321$

   (b) CRL $\text{Number}_{CRL\ D} = 333$
   Base CRL $\text{Number}_{CRL\ D} = 321$
   CRL $\text{Number}_{CRL\ E} = 333$

Table A.5 Revoked certificates

Issuer	Serial number	Revoked
CN = First CA, C = DE	1	True
CN = First CA, C = DE	2	True
CN = First CA, C = DE	3	False
CN = First CA, C = DE	4	False
CN = First CA, C = DE	5	False
CN = Second CA, C = DE	1	False
CN = Second CA, C = DE	2	False
CN = Second CA, C = DE	3	True
CN = Second CA, C = DE	4	True
CN = Second CA, C = DE	5	True
CN = Second CA, C = DE	6	False
CN = Third CA, C = DE	6	True
CN = Third CA, C = DE	7	False
CN = Forth CA, C = DE	6	False
CN = Forth CA, C = DE	7	True

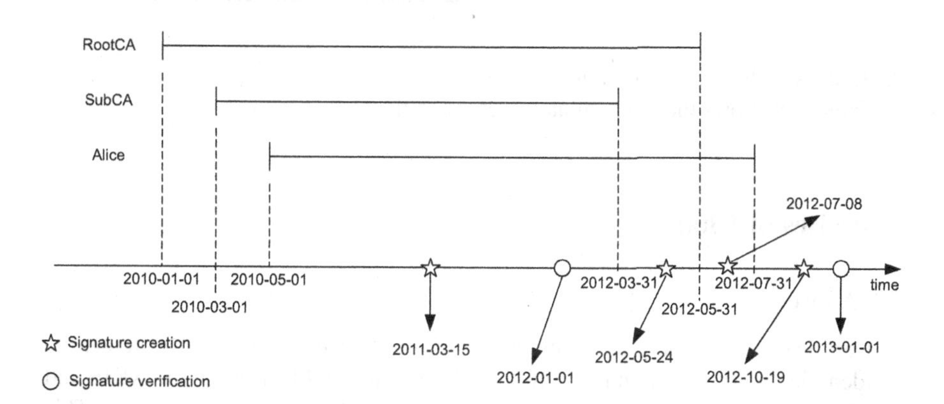

Fig. A.10 Signature and verification time points of Exercise 6.1

(c) CRL $\text{Number}_{CRL\,D} = 333$
Base CRL $\text{Number}_{CRL\,D} = 234$
CRL $\text{Number}_{CRL\,E} = 333$

5.4 The correct values are shown in Table A.5.

Exercises of Chap. 6

6.1 Figure A.10 shows the relevant dates on a time line.

1. Table A.6 shows the validation values.

Table A.6 Validation in different models

Signature creation time	Shell model	Modified shell model	Chain model
2010-04-20	Valid	Invalid	Invalid
2011-03-15	Valid	Valid	Valid

Table A.7 Validation for different signature times

Signature creation time	Shell model	Modified shell model	Chain model
2011-03-15	Invalid	Valid	Valid
2012-05-24	Invalid	Invalid	Valid
2012-07-08	Invalid	Invalid	Valid
2012-10-19	Invalid	Invalid	Invalid

Table A.8 Valid signature creation times

Model	From	Until
Shell	2010-05-01	2012-03-31
Modified shell	2010-05-01	2012-03-31
Chain	2010-05-01	2012-07-31

2. Table A.7 shows the validation values.
3. Table A.8 shows the valid signature creation times.

Exercises of Chap. 7

7.1 Answer:

1. Certificate 4 is a root certificate and a CA certificate. A root certificate can be identified by the fact that it is self-signed since the keyIdentifier of the AKIE and SKIE are the same. Certificate 5 is a CA certificate but not a root certificate. This can be seen from the basic constraints extension.
2. To find the public key of the issuer of certificate Z, a certificate must be found in which the subject is the issuer of Z. If there are several such certificates, they must all be tried. An alternative is to find a certificate in which the subject key identifier is equal to the authority key identifier of Z. A third possibility is to find a certificate in which the issuer and the serial number are equal to the authorityCertIssuer and authorityCertSerialNumber in the authority key identifier extension in certificate Z.
3. The key usage extension is marked critical. This means that the application must reject the certificate if it cannot interpret this extension.
4. The CA needs Bob's public key. PoP can be implemented by having Bob sign a challenge, typically the public key combined with a random number to prevent replay attacks. For example, this is done in PKCS#10.

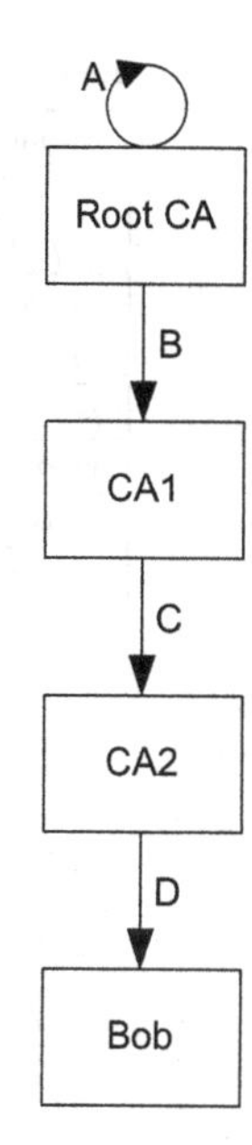

Fig. A.11 Certification hierarchy for the certificates of Exercise 9.1

Table A.9 Input variables of the algorithm

cp	Certificates B, C, and D
date	2012-03-23
uips	{ ANY }
public key	Key-0x77004433
ipolmap_inh	False
iexpol	True
iapol_inh	False
ipersub	{}
iexsub	{}

5. Since the key may only be used for encryption (key usage), signing cannot be used in the PoP. Now there are two options. The direct method: Upon receiving the public key, the registration authority sends an encrypted random number to the owner of the public key who decrypts it and sends it back. The indirect method: The RA accepts the public key without checking it. The CA generates the certificate and encrypts it with the public key. The participant can only use the certificate if he can decrypt it.
6. A certificate binds an identity to a public key. The CA must ensure that the corresponding key pair exists. For example, this may not be the case if the public key is changed because of transmission errors. Without PoP for signature keys the following attack is possible. The adversary asks for a certificate containing his own identity as the subject but the public key of another entity. Then, the signatures of the other entity will be interpreted as the signatures of the adversary.
7. This makes sense for Certificates 2 and 3. They contain encryption keys. A key backup allows encrypted data to be decrypted even if the private key is lost.

Table A.10 Variables of the algorithm and their values during execution

Variables:	Iteration			
	$i = 1$	$i = 2$	$i = 3$	(wrap-up)
v_p_t	See Fig. A.12			
ex_pol	0	0	0	0
in_ap	4	3	2	2
pol_map	4	3	2	2
w_pk	0x77004433	0xBBDD5588	0x12340987	0x80907060
w_iss	CN = RootCA	CN = CA1	CN = CA2	CN = CA2
m_path	3	1	0	0

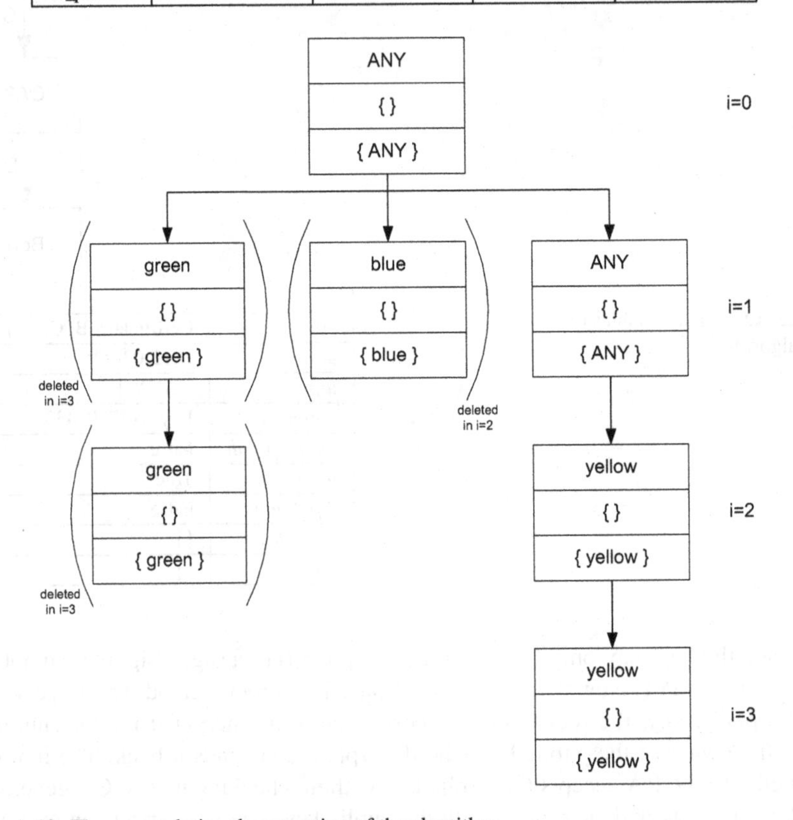

Fig. A.12 The *v_p_t* during the execution of the algorithm

Exercises of Chap. 8

8.1 Answer:

1. Certificate A maps policy magenta to policy red, but magenta is not listed in the certificate policies. Although this should not be used, the certificate is valid.

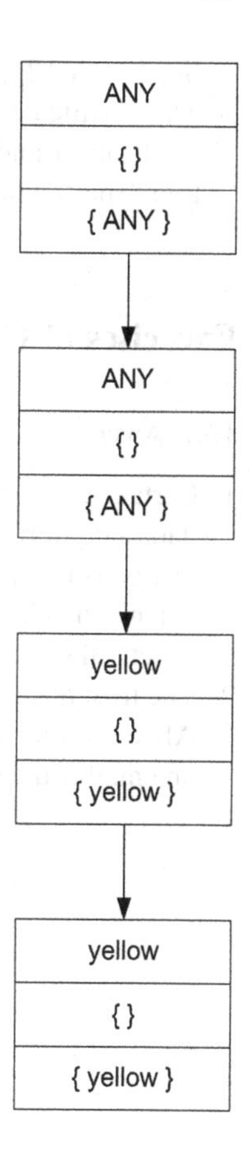

Fig. A.13 The v_p_t when the algorithm terminates

2. Certificate B is not valid. It contains a policy mappings extension but it is not a CA certificate.
3. Certificate C is valid.
4. Certificate D is not valid because it maps anyPolicy to policy yellow.

Exercises of Chap. 9

9.1 Answer:

1. See Fig. A.11.
2. The length of the certification path *cp* is $n = 3$. It contains the certificates B, C, and D. Certificate A is the self-signed certificate of the Root CA and is excluded from the path.
3. See Table A.9.
4. See Table A.10.

5. See Fig. A.12.
6. The certificate is valid. The output of the algorithm indicates the successful verification and contains the public key *key-0x80907060* and the valid policy tree depicted in Fig. A.13.

Exercises of Chap. 10

10.1 Answer:

1. Certain extensions of Bob's certificate do not contain appropriate values. The dataEncipherment value must be set in the key usage extension. The emailProtection (or anyExtendedKeyUsage) value must be set in the extended key usage extension. The rfc822Name field of the alternative name extension must contain exactly the same email address as the recipient field of the email.
2. The trust anchor to validate Bob's certificate is not installed in the email client.
3. Alice does not have an appropriate certificate. Many email clients refuse to send an email if the sender certificate does not permit encryption.

Index

J.A. Buchmann et al., *Introduction to Public Key Infrastructures*,
DOI 10.1007/978-3-642-40657-7, © Springer-Verlag Berlin Heidelberg 2013

C

Q

R

S

T

U

Printed in the United States
By Bookmasters